Popular Mechanics

WHAT WENT WRONG

Book Design by Naomi Mizusaki for supermarket

Library of Congress Cataloging-in-Publication Data

Hayes, William.
  What went wrong : investigating the worst man-made and natural disasters / William Hayes.
      p. cm.
  "Popular mechanics."
  Includes index.
  ISBN 978-1-58816-545-9
  1. Natural disasters–Juvenile literature. 2. Disasters–Juvenile literature. 3. Accidents–Juvenile literature. I. Title. II. Title: Popular mechanics what went wrong.
  GB5019.H39 2011
  904'.5–dc22

                          2010028947

10 9 8 7 6 5 4 3 2 1

Published by Hearst Books
A division of Sterling Publishing Co., Inc.
387 Park Avenue South, New York, NY 10016

Popular Mechanics is a registered trademark of Hearst Communications, Inc.

www.popularmechanics.com

For information about custom editions, special sales, premium and corporate purchases, please contact Sterling Special Sales Department at 800-805-5489 or specialsales@sterlingpublishing.com.

Distributed in Canada by Sterling Publishing
c/o Canadian Manda Group, 165 Dufferin Street
Toronto, Ontario, Canada M6K 3H6

Distributed in Australia by Capricorn Link (Australia) Pty. Ltd.
P.O. Box 704, Windsor, NSW 2756 Australia

Printed in USA

Sterling ISBN 978-1-58816-545-9

PopularMechanics

WHAT WENT WRONG

INVESTIGATING THE WORST MAN-MADE AND NATURAL DISASTERS

WILLIAM HAYES

HEARST BOOKS

A division of Sterling Publishing Co., Inc.

New York / London

www.sterlingpublishing.com

CONTENTS

FOREWORD

A jet flying from Rio de Janeiro to Paris vanishes without a trace. Insufficient preparation for an earthquake devastates Port-au-Prince, Haiti. A chain of poor decisions leads to a blowout on the Deepwater Horizon semi-submersible drilling rig, releasing a flood of oil into the Gulf of Mexico.

Over the past century, *Popular Mechanics* has investigated the world's worst natural and manmade disasters. While their effects are devastating, each event offers the possibility for engineers, scientists, and leaders in business and government to compel technological advances, learn from their past mistakes, and, most important, prevent future loss of lives. Here you'll discover breakthroughs in science, as well as individual acts of heroism, that occurred at some of humanity's low points.

After the earthquake in the Indian Ocean triggered a tsunami in December 2004, the United Nations started work on the Indian Ocean Tsunami Warning System, which was completed in 2006. The flooding in New Orleans after Hurricane Katrina spurred FEMA to require all levees to carry a 100-year flood mark certification, a robust standard that requires the entire system to withstand a storm the likes of which we see no more than once every hundred years. The aviation industry rigorously studies every accident, leading to the development of improved safety features; the Federal Aviation Administration reports only one fatal crash between 2002 and 2007, an astounding number considering that more than thirty thousand flights take off every day.

We've documented the improvements and safety measures that were developed in the wake of disaster, some of which can be readily adopted by any of us. Yet today, only 12 percent of Americans have taken adequate measures to prepare for crises. Are you one of them? If you're not, whether you endure a house fire or a hurricane, this book provides personal, detailed information to help protect your family and home before calamity strikes.

The Editors of *Popular Mechanics*

PART

1

NATURE'S FURY

EARTHQUAKES

Earthquakes occur when massive pieces of the earth's crust, called plates, crash into, over, under, and across each other. The large continental plates upon which the land and the sea rest move just millimeters per year, or about the speed at which the average human fingernail grows. These minute speeds and distances can cause widespread destruction around and near the plate's boundaries.

Geologists measure the seismic energy that earthquakes release with the Richter scale. Smaller earthquakes—below 5 on the Richter scale—happen daily, but most are too weak, too deep underground, or too far from a population center to feel. An earthquake above 9.5 has never been recorded, although some have come close.

← A man looks for items to scavenge from the rubble in the Haitian capital of Port-au-Prince on January 27, 2010. Six days earlier, a quake that reached the cataclysmic magnitude of 7.0 on the Richter scale had its epicenter just 20 miles from the city. Dozens of aftershocks rocked the country in the following weeks, and an estimated quarter million people perished. Haiti was already the Western Hemisphere's poorest country, and this disaster plunged the country deeper into poverty, collapsing hundreds of houses and leaving more than one million citizens homeless.

On February 27, 2010, one of the ten most powerful earthquakes in recorded history struck off the coast of central Chile. It sent tsunamis slamming into coastlines as far away as Japan. Although this quake was around six hundred times more powerful than the 7.0 earthquake in Haiti that had occurred the month before, the Chilean quake did not garner nearly as much attention. In Haiti, however, aid continued to pour in, news reports continued to pour out, and the death toll continued to rise. What saved Chile?

Chileans have long been familiar with earthquakes. The most powerful earthquake ever recorded, a 9.5, occurred in Valdivia, Chile, in 1960. The 2010 temblor was the seventh in Chile that measured 7.0 or greater since the Valdivia quake. Most Chileans have therefore grown up with earthquakes; they are a fact of life, like the country's mountainous terrain and beautiful scenery. Buildings are routinely built with earthquake resistance in mind, and the Chilean government holds architects and contractors to strict building codes and standards.

Another factor that mitigated the damage was the location of the earthquake's epicenter: about 75 miles northwest of Concepción, Chile's second largest city, and 200 miles southwest of the capital city of Santiago. The epicenter occurred about 60 miles out to sea and more than 20 miles underground as well, so much of its seismic energy had dissipated by the time it reached the surface.

The Aftermath

Most of the damage caused by the earthquake was centered around Concepción, where looters added to the destruction caused by the quake. The temblor killed about seven hundred people and

caused roughly $30 billion of damage. It sent tsunamis around the world, to places as far away as Antarctica and Alaska. Coastal areas in Hawaii were evacuated, but the waves were smaller than expected and did not cause much damage. No casualties were reported in Japan, the farthest country reached by the tsunamis.

A culture of preparedness helped save Chile. Many earthquake-prone areas would do well to take a lesson from the Chileans. But when earthquakes strike areas with less experience in dealing with them, the results can be devastating.

A destroyed building in Concepción, Chile, February 27, 2010. The earthquake rated a magnitude of 8.8 on the moment magnitude scale and lasted a full 90 seconds. It was strongly felt from the Chilean region of Valparaiso in the north, to Araucania in the south. Because the epicenter was located some 60 miles out to sea, much of the quake's seismic energy had dissipated.

The Chilean earthquake triggered a tsunami that devastated several coastal towns in south-central Chile, including Dichato Beach (pictured). The earthquake caused a blackout that killed power in more than 90 percent of the country. Despite the damage, the death toll from the quake was miniscule in comparison to the Haiti earthquake just one month earlier.

HELL IN HAITI

January 12, 2010: Fault Lines Pointed to Disaster

On January 12, 2010, a 7.0-magnitude earthquake struck Haiti, decimating the island nation. As government officials and rescue agencies were sorting through the rubble, many people had the same thought: could this tragedy have been prevented?

Some scientists thought so. Back in 2008, Eric Calais and Paul Mann, geophysicists who study fault lines in the Caribbean, had predicted that Haiti would soon face such a devastating quake. The researchers reported that the Enriquillo fault, the line that Haiti sits upon, could produce a 7.2-magnitude quake if strained enough. Using global positional system (GPS) measurements, the team calculated that the fault was inching along at 7 millimeters per year, a moderate crawl in the

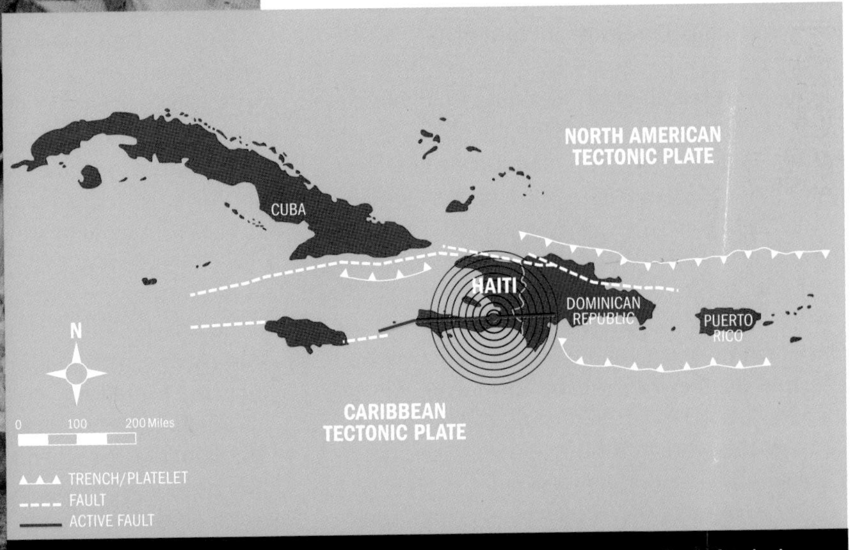

NORTH AMERICAN TECTONIC PLATE

CUBA

HAITI

DOMINICAN REPUBLIC

PUERTO RICO

CARIBBEAN TECTONIC PLATE

N

0   100   200 Miles

▲▲▲ TRENCH/PLATELET
- - - - FAULT
──── ACTIVE FAULT

A woman looks down a Haitian street following the Haiti earthquake. The Caribbean tectonic plate shifts a quarter of an inch a year in relation to the North

American plate. Much of the architecture in Haiti, such as the buildings on this street, was unable to withstand the power of an earthquake that measured 7.0 in magnitude.

Geologists had warned that there was a high risk of significant seismic activity in Haiti back in 2008, and had recommended "high priority" rupture studies.

Haiti failed to implement emergency plans and restructure crucial buildings.

realm of fault lines. But this fault line has stretched several millimeters per year for the last 250 years, and it was time for it to snap.

"Unfortunately, our prediction was fairly close to what actually happened. Think of the fault as a rubber band being pulled 7 millimeters per year at a constant rate. It will eventually break," Calais said.

Mann equated the fault in Haiti with the San Andreas fault in California, as both have plates that slip and grind past one another in a horizontal direction. The difference between the two is that Haiti's hasn't been quantitatively studied in the past. Because Haiti is a difficult work environment that poses safety concerns and has a poor highway system, it has been neglected by seismologists, according to Calais. Fortunately, by measuring the speed of the Enriquillo fault line, his team has made substantial progress in Caribbean geophysics.

But his research didn't translate well enough to elicit safety precautions before the quake. Though earthquakes can't be prevented, Calais noted that there was enough advance warning for the Haitian government to have made preparations, and, in fact, his team had alerted the government four to five years beforehand.

"We told the Haitian government that the Enriquillo fault is a major player," Calais said. "We told them exactly where the fault is. We told them how fast it was building up elastic energy, and that if it were to rupture, it could produce a 7.2-magnitude—or larger—event."

The government worked with the team and listened to its foreboding reports, but for the most part, Haiti failed to implement emergency plans and restructure crucial buildings. Even with scarce resources, there were options, according to Calais. For example, "You can identify the few buildings that are critical, that have to stand up in the face of a large earthquake, like hospitals and schools, from which rescue operations can be organized. This hasn't been done," he said. "One of the first buildings that was reported to have collapsed was a hospital. That is unacceptable and could have been prevented."

The Aftermath

What was already the Western Hemisphere's poorest country plunged further into desperation. More than two hundred thousand were estimated dead, and more than a million were left homeless. Hundreds of thousands of poorly constructed houses collapsed around their unfortunate occupants in the capital city, Port-au-Prince.

Port-au-Prince suffered so severely due to its close proximity to the epicenter, about 20 miles west. The earthquake also occurred not far beneath the earth's surface—only 6 to 7 miles down. The poverty in which many Haitians live contributed further to the death toll. Rickety wooden house frames crumpled like paper under the weight of their hurricane-resistant concrete roofs. Building codes in Haiti are virtually nonexistent, so many buildings are thrown together as quickly, easily, and cheaply as possible, with little regard to structural integrity.

Calais hopes that neighboring areas such as the Dominican Republic take note of the Haitian quake and heed the dangers fault lines forewarn.

"If countries don't learn from this and make preparations, it is very sad," he said. "Hopefully something positive will come out of this tragedy in Haiti, that people will learn the hazards of earthquakes."

ANATOMY OF A SHAKEMAP

Over the past decade the amount of information that scientists have obtained from earthquakes has increased dramatically, while that shared with the public has remained pretty much unchanged. After a quake occurs, seismic laboratories typically issue two pieces of information. The first is the Richter reading, which defines an earthquake's intensity based on ground motion. The greater the number, the greater an earthquake's energy release. Seismologists also identify the epicenter, the location where the Earth first moved. Neither piece of information explains why severe damage can occur tens of miles from the epicenter.

This is where ShakeMaps come in. Within minutes, they provide an easy-to-comprehend diagram that shows an earthquake's local impact. In the ShakeMap below, sections highlighted in red represent areas that sit atop geological structures that reacted most violently to ground movement. When people are trapped in collapsing buildings, surrounded by popping gaslines and snapping powerlines, time is of the essence. ShakeMaps provide the type of information emergency service managers need to decide how to deploy rescue and medical services.

Color is a useful way to present complex information to the human mind. This information also can be conveyed as strings of numbers and be used to create a dynamic model of a geological region inside a high-speed computer. The conditions within this mathematical model are altered by changes in the real world, communicated through a network of in-ground sensors that are strung through earthquake-prone regions. In much the same way that a glass will produce different pitches depending upon the level to which it is filled, the mathematical model reacts differently depending upon which sensors are activated.

As useful as ShakeMaps will be for emergency workers, they cannot match the lifesaving potential of knowing where and when the next earthquake will hit. Our current ability to answer this question is so vague as to be useless. For example, everyone knows that Southern California is overdue for "the big one." Predictions based on observing signaling activities, such as changes in patterns of smaller quakes, have met with only limited success, according to the National Academy of Science (NAS).

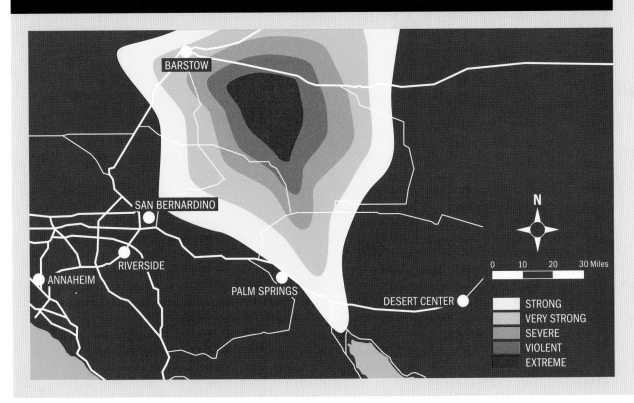

SAN FRANCISCO

The Great California Quake of April 18, 1906

As terrifying as the rumbling of the earth is, the effects and aftereffects of earthquakes are usually more dangerous than the quakes themselves. Buildings collapse, trapping people. Gas lines rupture and fire rips through affected areas. Frightened, hungry people riot. The San Francisco earthquake of Wednesday, April 18, 1906, is one case in which the quake's aftermath was more devastating than the shaking of the earth.

San Francisco rose from the gold rush of 1849, which transformed the city from a rough-and-tumble mining town into a cosmopolitan center of four hundred thousand people. Aspirational locals called it the Paris of the West. There were fashionable department stores, urbane hotels, a new sprawling city hall said to be the biggest in the West, and the Grand Opera House, which hosted the greatest tenor of the time, Enrico Caruso. All of that changed in seconds.

At 5:12 AM, a powerful earthquake centered just off the coast grabbed San Francisco by the throat and nearly shook it to death. The magnitude 7.8 quake arrived in two pulses, the second more powerful than the first.

Ship captains said it felt as though their boats had run into a sea of rocks. Wooden houses splintered, cracked, and collapsed, while poorly reinforced brick buildings tumbled to the ground. Bleary residents scurried into streets that were rippling like waves and firing off cobblestones. Trees whipsawed, telephone poles snapped, and streetcar rails buckled.

The epicenter of the earthquake was 2 miles out to sea on the San Andreas fault, but the focal point of the quake's damage was in the working-class neighborhood south of Market Street. When the shock waves rippled through this reclaimed swampland, they temporarily liquefied the man-made ground, causing scores of buildings to

The epicenter of the quake occurred offshore just two miles from the city, near Mussel Rock. Aftermaths were felt in downtown San Francisco, including Market Street. The heavily damaged Phelan Building stands in the foreground. The earthquake's aftermaths were more destructive than the powerful quakes themselves.

The city was in ruins.

Sacramento Street in the Mission District burns. Despite the damage of the earthquake and aftershocks, 90 percent of the city's destruction was the due to the subsequent fires. More than 30 separate fires destroyed 25,000 buildings. The resultant destruction and chaos led to widespread looting throughout the city, and the mayor was forced to issue an illegal shoot-to-kill order to staunch the rampant lawlessness.

collapse. Chinatown, just north of Market Street, was also particularly hard hit because of the extensive use of unreinforced brick masonry in its buildings.

The quake ruptured gas lines, sparking dozens of fires to life; they caused yet another, more powerful rash of destruction. The tremors had broken the city's fire alarm system, but firefighters could see billowing smoke and knew where to go. They hooked fire hoses to hydrants, but most of the city's water lines had ruptured, too: there was no water.

The tightly packed wooden-frame construction concentrated south of Market Street made fast fuel for a blaze that jumped from building to building and from block to block. The fire completely destroyed a local college, San Francisco's Hall of Records, and the massive city hall.

The fate of those still trapped by earthquake debris was much worse. At the collapsed Valencia Street Hotel, rescuers dug feverishly to free survivors but were forced to retreat as the fire descended. An estimated one hundred people didn't make it out of the rubble.

The city's fire chief, Dennis Sullivan, appreciated the hazards of a city with tightly clustered wooden buildings, and in the year before the earthquake had spoken of dynamiting buildings to create firebreaks in the event of massive fires. But the chief was mortally injured in the first minutes of the 1906 quake, and with him went any semblance of a plan.

When the shock waves rippled through this reclaimed swampland, they temporarily liquefied the man-made ground, causing scores of buildings to collapse.

His successor, John Dougherty, had no expertise with dynamite. He contacted the army base at the Presidio and asked for the help of General Frederick Funston, the brigadier general in charge.

The dynamiting doomed much of the city that wasn't already burned, causing its own fires and refueling others. In Chinatown, an estimated sixty fires were started this way. Flaming debris from buildings ignited natural gas from ruptured lines, and fire engulfed neighboring buildings.

The next day the fire—and the dynamiting—continued. On Friday, with the core of the city smoldering, firefighters mounted a final campaign of total destruction to hold the fire at Van Ness Avenue. The firebreak worked. But the city was in ruins. According to the *San Francisco Chronicle*, "522 city blocks, 4 square miles…2,593 acres, [and] 28,188 buildings" were utterly destroyed.

Looting in the days after the quake was said to be rampant. The mayor issued an illegal shoot-to-kill order to staunch the looting, and the army pressed citizens into work crews at gunpoint. The city's leaders reported the official death toll to be less than five hundred, but historians have since argued that anywhere from three to six thousand people were killed. Some three hundred thousand evacuated by ferry and train.

The Aftermath

Within days of the quake, Governor George Pardee of California commissioned an exhaustive earthquake investigation and appointed Andrew Lawson, a geology professor at the University of California, Berkeley, to lead it. In what is now commonly referred to as the Lawson Report, the investigators mapped nearly 300 miles of ruptured fault line and thoroughly documented structural damage in relation to local geology and shaking intensity.

But the hows and whys of the 1906 earthquake were still not well understood. Drawing on observations collected in the report, Henry Fielding Reid, a geology professor at Johns Hopkins University, later proposed the theory of elastic rebound, in which pressure in the earth's crust is built up and then released like a snapped rubber band. Reid's theory led to the all-inclusive theory of plate tectonics, which wouldn't gain credibility for another fifty years, but Reid "had it all right," said Mary Lou Zoback, a geologist with the U.S. Geological Survey. The investigation is the foundation for all earthquake research that followed, Zoback adds, and it is still used today.

10 MOST POWERFUL RECORDED EARTHQUAKES

Location	Magnitude	Year
Valdivia, Chile	9.5	1960
Prince William Sound, Alaska	9.2	1964
Off the coast of Sumatra, Indonesia	9.1	2004
Kamchatka	9.0	1952
Off the coast of Ecuador	8.8	1906
Chile, Maule Region	8.8	2010
Rat Islands, Alaska	8.7	1965
Sumatra, Indonesia	8.6	2005
Tibet, Assam	8.6	1950
Andreanof Islands, Alaska	8.6	1957

ANATOMY OF
AN EARTHQUAKE

Plate Tectonics is the scientific theory of large-scale motions of the Earth's lithosphere. It is widely believed that there are seven major tectonic plates resting on the asthenosphere, and several other minor ones. The movement of these plates, while barely perceptible, is capable of producing some of the most dramatic and harrowing images of human suffering in the world today.

Earthquakes occur when the energy built up in the friction created between the plates is discharged. The result is a massive release of energy that creates seismic waves of elastic energy below the Earth's surface, causing earthquakes, volcanic activity, oceanic trench and mountain building, and island arcs.

Tectonic plates move at about the speed at which a hair or fingernail grows, but this tiny change can cause a tremendous amount of devastation. Whether it's a volcano spewing clouds of ash or a tsunami threatening a coastal city, it happens because of earthquakes caused by plate motions at one of three different boundaries:

- **Convergent Plate Boundary:** These plates move toward each other, creating reverse faults. These can either be a Subduction Boundary, where oceanic lithosphere is pushed below continental lithosphere, or a Collision Boundary, where plates with continental lithospehere smash together. Both create mountain ranges. When two oceanic lithospheres collide, they create island arcs above.

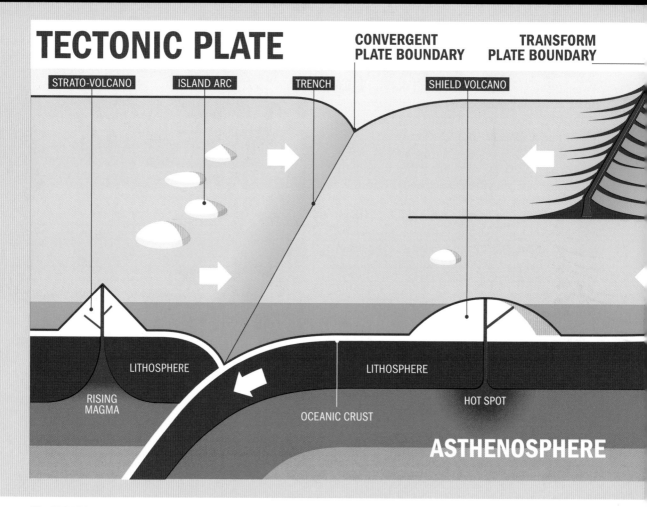

TECTONIC PLATE **CONVERGENT PLATE BOUNDARY** **TRANSFORM PLATE BOUNDARY**

STRATO-VOLCANO   ISLAND ARC   TRENCH   SHIELD VOLCANO

LITHOSPHERE   LITHOSPHERE

RISING MAGMA

OCEANIC CRUST

HOT SPOT

ASTHENOSPHERE

- **Divergent Plate Boundary:** This is a zone where two plates, usually at an oceanic ridge at the lithosphere, move in opposite directions. The stress on the subducting plates tends to cause shallow-focus earthquakes with depths less than 12 miles, and usually creates a series of volcanic islands.

- **Transform Fault Boundary:** When plates slide past each other horizontally, such as the San Andreas Fault in California, they produce earthquakes with focal depths less than approximately 35 miles.

Approximately two dozen oceanic trenches exist on the floors of the earth's oceans, most of which are located in the Pacific. These trenches are responsible for some of the most violent earthquakes in history and for sending incredibly deadly tsunamis surging toward coastal towns and cities.

This Pacific Ring of Fire, a horseshoe-shaped arc, is home to nearly 500 volcanoes and three-fourths of the active and dormant volcanoes on earth. Throughout the Aleutians, Japan, and the Philippines, there are clusters of island arcs which border the Pacific Plate.

Because of plate tectonics, cold and wet oceanic crust drops down into the asthenosphere at convergent plate boundaries. This action causes the subducting plate to melt, creating magma that can rise through strato volcanoes. At its most potent, these volcanoes spew the burning-hot lava that has terrorized man since ancient times.

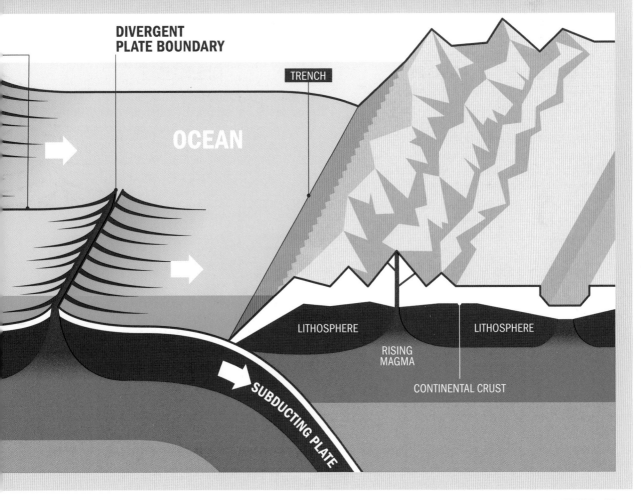

DIVERGENT PLATE BOUNDARY

TRENCH

OCEAN

LITHOSPHERE

LITHOSPHERE

RISING MAGMA

CONTINENTAL CRUST

SUBDUCTING PLATE

It is clear what havoc an earthquake and its aftermath can cause in a bustling city. San Francisco's 7.8-magnitude quake was mighty, but not enough to make it one of the ten most powerful in recorded history. At 5:36 PM on March 27, 1964, the largest earthquake ever recorded in North America—and the second largest in history—rattled coastal Alaska for close to 4 minutes. Though the epicenter of the Great Alaskan Earthquake was deep beneath Prince William Sound—75 miles east of Anchorage and 56 miles west of Valdez—the magnitude 9.2 temblor rippled water as far away as Louisiana and even made parts of Florida and Texas jump a couple of inches.

Every year the Pacific tectonic plate barges roughly 2 inches into, and under, the North American plate near southern Alaska. Under intense pressure from the friction, the earth's crust bends and strains until it eventually snaps back into place, as it did that March evening. In downtown Anchorage, the quake caused the streets to become asphalt waves, bouncing cars into the air. Parts of the city dropped as much as 30 feet, bringing the Denali Theater marquee on Fourth Avenue to sidewalk level. One block over, the concrete facade of the new J. C. Penney crashed into the sidewalk, killing a crouching pedestrian and a passing driver. At the airport, the control tower toppled over and killed an air traffic controller.

The ground in the Turnagain Heights subdivision simultaneously sank and surged up. "Our whole lawn broke up into chunks of dirt, rock, snow, and ice," Turnagain resident Tay Pryor Thomas wrote in *National Geographic* shortly after the quake. "We were left on a wildly bucking slab; suddenly it tilted sharply, and we had to hang on to keep from slipping into a yawning chasm."

But what claimed 115 of the 131 lives that day wasn't the earthquake itself. It was the tsunami waves that screamed across Prince William Sound and down the Pacific Ocean. The quake had caused several underwater landslides that, in turn, displaced vast amounts of water. The great volume of water that was forced out to sea returned just as quickly, in the form of giant waves that geologists call local tsunamis.

"I was in my house in Cordova eating dinner when the quake struck," recalls Pete Corson, then a U.S. Coast Guard lieutenant assigned to the cutter *Sedge*. "Our house came almost completely off the foundation. I ran down to the dock, and it had split in half and was heaving back and forth. We had to wait until the gap closed before jumping across it to get to the *Sedge*." Corson brought his wife and three young sons with him and boarded the cutter. Soon, reports began to crackle over the ship's radio: "The village of Chenega is completely wiped out by a tidal wave. There are many injuries…[and] only one house left standing." The *Sedge* was in "Charlie status," with its engines disassembled, and the crew had to scramble to get her underway.

"We had to hang on to keep from slipping into a yawning chasm."

A landslide damaged a number of houses along L Street. The Four Seasons Apartment Building collapsed, and beyond it stands an undamaged three-story reinforced concrete frame building on a more stable block. The 9.2-magnitude quake rippled water as far away as Louisiana and was felt as far away as Florida.

Fissures in Seward Highway appeared near the Alaska Railroad station at Portage, following the Great Alaskan Earthquake. At some places, tectonic subsidence and consolidation of alluvial materials dropped both the highway and the railroad below high-tide levels. Parts of the city of Anchorage dropped as much as 30 feet during the earthquake.

Forty-five miles north of Cordova, in Valdez, the 400-foot freighter SS *Chena* was in port when the quake hit. Twenty seconds after the initial tremors, 98 million cubic yards of the silty delta slumped away. The massive amount of displaced water came rushing back as a 40-foot local tsunami, killing more than thirty people as they tried to flee.

"I saw people running—it was just ghastly," said Merrill Stewart, captain of the *Chena*, which keeled over as the water ran out of the inlet. "They were just engulfed by buildings, water, mud, and everything." The *Chena*, remarkably, was able to right itself and radio this message: "The town of Valdez, Alaska, just burst into flames. The whole dock is afire…."

In nearby Shoup Bay, the local tsunami reached 200 feet in height and destroyed a remote cabin. Thirteen people were killed in Whittier as three local tsunamis, the last one 42 feet high, crashed into town. And in Chenega, an 89-foot wave killed twenty-three of the town's seventy-five residents.

Similar stories unfolded along Prince William Sound. In Seward, the fuel tanker *Alaska Standard* was in port loading gasoline and diesel when the quake hit. The dock caught fire as the harbor shore area slumped into the bay. When the water returned, a 30-foot tsunami destroyed a fuel tank farm, where oil is stored before being offloaded, setting it ablaze. As the fire spread, more than three dozen fuel-loaded tanks exploded.

A tectonic tsunami, different from the landslide-generated local tsunamis, was caused as the quake heaved 100,000 square miles of Prince William

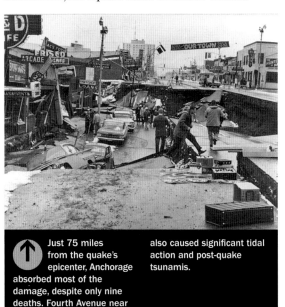

Just 75 miles from the quake's epicenter, Anchorage absorbed most of the damage, despite only nine deaths. Fourth Avenue near C Street collapsed after the landslide. The earthquake also caused significant tidal action and post-quake tsunamis.

Sound either up or down. Near Kodiak, the ground surged permanently out of the water by 30 feet, while near Portage it dropped by 8 feet. The displaced water went pulsing down and across the Pacific Ocean and didn't stop until it hit Japan.

The local tsunamis struck within seconds; it took longer for the tectonic tsunami to radiate out. Twenty minutes after the local wave set Seward on fire, the tectonic tsunami rolled in looking as if it, too, was aflame. "[As] the fire was really roaring, the wave came up Resurrection Bay and spread it everywhere. It was an eerie thing to see—a huge tide of fire washing ashore," recalled railroad employee Gene Kirkpatrick to *National Geographic* a few days later.

Back in Cordova, which had suffered only minor damage, the crew of the *Sedge* finally got underway and was ordered to Valdez. The cutter was in Cordova's 60-foot shipping channel when it began to drop.

"Our Fathometer just kept dropping," remembers Pete Corson, who realized the tide was being sucked out of the channel. "It was very strange. It was incredibly dark outside, and giant snowflakes were falling, so it was very hard to see, but we could hear loud clapping—which turned out to be halibut flapping on the seabed." The *Sedge* came to rest on the channel bottom, and 10 minutes later the water rushed back in. "We rode it out under anchor and then went on our way."

The tsunami continued its inexorable southerly push at 415 mph. For the next 2 hours it rolled down the coastline of British Columbia, damaging a logging camp in Shield's Bay and destroying sixteen houses in Hot Springs Cove.

In Newport, Oregon, around 11 PM, it came ashore at Beverly Beach State Park, where a family of five was camping. The mother and father survived, but all three children were swept away. At 11:52, the first of four tsunami waves struck Crescent City, California, just south of the Oregon border. All ten deaths, and nearly all of the damage, were a result of the final 15.7-foot wave that struck at 1:45 AM.

"It was like a violent explosion. A thunderous roar mingled with all the confusion. Everywhere we looked, buildings, cars, lumber, and boats shifted around like crazy. The whole beachfront moved, changing before our very eyes. By this time, the fire had spread to the Texaco bulk tanks. They started exploding one after another, lighting up the sky," recalled Peggy Coons in *The Raging Sea*, a book about the quake by Dennis Powers.

By the time it got to San Francisco, the wave was just a couple of feet high. It damaged several yachts and did more of the same in Hilo, Hawaii. When it reached Japan it was barely visible, but it was still strong enough to damage pearl farms along the country's eastern coast.

Aftermath

The Great Alaskan Earthquake and tsunami of 1964 caused more than $300 million in damage along the Pacific Coast, from Anchorage to Los Angeles, according to a report compiled by the West Coast and Alaska Tsunami Warning Center. In fact, the 1964 tsunami was responsible for the creation of the center, located in Palmer, Alaska. Since 1967, scientists on round-the-clock duty monitor seismic activity, tidal gauges, and data collection buoys in order to determine when to issue tsunami warnings. Besides being broadcast to the public, the warnings also trigger alerts to local, state, and federal emergency officials, as well as to the military and the Coast Guard.

After the quake, the state of Alaska and the federal government went to work cleaning up. The U.S. Army Corps of Engineers spent $110 million dollars rebuilding roads and clearing debris in Alaska. The native village of Chenega, which had been completely destroyed, was moved to higher ground. Likewise, the town of Valdez, which sat at the mouth of a silty glacial drainage, was abandoned and rebuilt a few miles west atop a foundation of solid bedrock. The structural damage to downtown Anchorage was extensive, but the timing of the earthquake, Good Friday evening, undoubtedly spared many lives.

As part of its Advanced National Seismic System, the U.S. Geological Survey has recently outfitted a few buildings with complex motion sensors in order to understand how they respond to quakes. Because of Anchorage's seismically active location (five to six thousand earthquakes occur in Alaska annually), its twenty-story Robert Atwood Building is one of the most thoroughly monitored structures in the country. The building is outfitted with 32 instruments from basement to ceiling to detect the swaying, twisting, rocking, and drift that result from seismic waves. Seven boreholes, ranging from the surface to a depth of 200 feet, also contain instruments that monitor seismic activity.

"We want to see what designs work, and what kind of damage, often structurally hidden, appears," says Mehmet Celebi, a research engineer with the *U.S. Geological Survey*. "In…two years… we've had twenty small- to medium-size earthquakes in the area. We're waiting for a large event to see how the building really reacts."

It is not yet possible to predict exactly when and where an earthquake will strike, but it may be in the near future. Technology such as ShakeMaps is facilitating faster, more accurate predictions. See page 17 for more details on the evolving field of ShakeMaps and earthquake detection.

With buildings collapsing, and jagged glass and hunks of concrete raining down, it's easy to get injured during an earthquake. Here's what to do if you're wounded by something heavy or sharp.

➡ Apply direct pressure by clamping your hand on the wound, then elevate the injury above your heart to slow the blood flow. If the bleeding continues for thirty minutes, use clothes to wrap the site in a pressure dressing.

➡ "Don't keep checking to see if [the dressing] is working, even if it's bloody," says Dr. Jeff Gutterman, a fellow of the American College of Emergency Physicians. "That's a classic mistake."

➡ If the bleeding doesn't stop after another 30 minutes, tie off the wound a few inches above the site. If you get emergency help within several hours, you probably won't lose the limb.

⬅ This house in Turnagain Heights, Anchorage, was severely and structurally damaged in the Alaskan Earthquake on March 27, 1964, far from the quake's epicenter. While over the past decade the amount of information that scientists can obtain from earthquakes has increased dramatically, the amount that is shared with the public has remained pretty much unchanged. After a quake occurs, seismic laboratories typically issue two pieces of information. The first is the Richter reading, which defines an earthquake's intensity based on ground motion. The greater the number, the greater an earthquake's energy release. Seismologists also identify the epicenter, the location where the earth first moved. Neither piece of information explains why severe damage can occur tens of miles from the epicenter.

VOLCANOES

The ancients rightly viewed volcanic eruptions with a mixture of awe and terror. They sought to rationalize such devastating phenomena—and who but the gods could make mountains explode? But their real cause is grounded in geology, or more specifically, vulcanology.

Small fissures in the earth's crust, often along fault lines, allow molten rock (magma) and gases to emerge from the asthenosphere. There are many different kinds of volcanoes, and different ways to classify them: by type and feature of the dome or mountain (shield volcano, cryptodome, stratovolcano, etc.), by activity level (active, dormant, extinct), and even by the nature of lava they eject. But the raw ingredients for volcanoes—magma, gas, and pressure—always have the potential for spectacular and deadly effects.

One of the most iconic American volcanoes is Mount St. Helens. Its 1980 eruption, which claimed fifty-seven lives, is etched in the mind of a nation. But as terrifying as that event was, it also spurred advances in volcanic study and eruption detection, and today it is far less likely that a volcano will catch Americans by surprise.

← Volcanoes can be things of visual beauty. Visible from miles away, even in the dark of night, volcanic activity is often awe-inspiring. But it is also capable of deadly havoc and economic disruption, and the hazards of volcanoes should not be taken lightly. Here, steam and lava pour out of Fimmvörðuháls, a ridge between Eyjafjallajökull glacier and Mýrdalsjökull glacier in Iceland.

On May 18, 1980, U.S. Geological Survey vulcanologist David Johnston had a clear view of Mount St. Helens's north flank from his monitoring station 5.5 miles away. Just seconds after the thirty-year-old radioed his final words to colleagues, the snowcapped volcano blew itself spectacularly apart.

The mountain had been tranquil until two months earlier, when an earthquake jolted the sleeping giant to life. A steady succession of more than ten thousand earthquakes and hundreds of explosive blasts of steam signaled the pressure building beneath the volcano's slopes. But by Saturday, May 17, the mountain seemed silent. The Red Cross had recalled its emergency cots, and many sightseers, having given up on a grand finale, had already gone home.

The next morning, at 8:32, Mount St. Helens began to tremble. What Johnston then witnessed, as he called in his warning, was the largest landslide in recorded history. A magnitude 5.1 earthquake cut loose the volcano's north flank, which had bulged by 450 feet. A wave of earth and ice rushed down the mountain at 150 mph, pouring into Spirit Lake and coursing 13 miles down the North Fork of the Toutle River.

Without a cap of earth to keep it sealed under pressure, Mount St. Helens exploded. Sulfur dioxide gas in the freshly exposed magma, together with compressed water and steam, expanded and blew out the north side of the mountain. This lateral blast of ash, magma, rocks, and sand reached 100 stories high and spread 10 miles wide as it plowed down valleys and over ridges at speeds near 700 mph.

It was this lateral blast, or surge, as some geologists prefer to call it, that caused most of the fifty-seven fatalities that day—including Johnston. Neither he nor his trailer was ever found; both were presumably hurtled into the next valley and buried in debris.

Most loggers were off for the weekend, but Jim Scymanky and three coworkers were roughly 10 miles away at the time of the blast. "It sounded like a couple of big passenger jetliners coming through the woods," Scymanky said at a Mount St. Helens commemorative event in 2000. "Everything was hot. Trees were smoldering." Scymanky's gloves had melted onto his hands, and the ash left serious burns over nearly half of his body. And then everything was gone.

"It was barren. You didn't know where the hell you were," Scymanky told the *Oregonian*. "One minute you've got landmarks all over. You know where the stream is, where the roads are. The next second there were no roads, no streams—no nothing.... It was like the blast picked you up and put you on a different planet." A National Guard helicopter rescued Scymanky and two of his coworkers, Leonty Skorohodoff and Jose Dias, who both later died from their wounds. The body of the fourth logger, Evlanty Sharipoff, was not found for more than a month.

"It sounded like a couple of big passenger jetliners coming through the woods."

The catastrophic eruption of Mt. St. Helens in Washington State was the deadliest and most economically destructive volcanic activity in U.S. history. Fifty-seven people were killed and thousands of big game animals were lost as the stratovolcano blasted a gigantic plume of ash that reached for miles around the mountain.

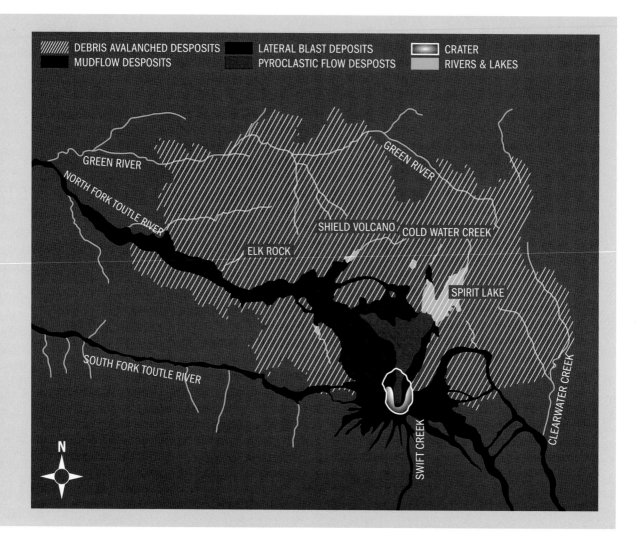

A few miles farther north, Bruce Nelson and a group of friends were fishing for steelhead trout on the Green River. Nelson's group, however, heard nothing as the surge closed in on them. Nelson later explained to the *Juneau Empire* how he survived. "Two huge trees were uprooted where my girlfriend, Sue Ruff, and I were standing. We fell down into the hole. That actually put us below ground level, which kept us from being burned severely. Our hair was frizzed, and the hair on our arms burned off, but we were saved from worse burns."

Nelson and his girlfriend trudged through knee-deep ash for hours until rescuers picked them up. They were among the estimated 125 people who survived in the blast zone, as were two of their friends. A third couple, Terry Crall and Karen Varner, died in their tent as the superheated ash blew through.

As far as 17 miles from the volcano, the surge flattened thousands of 100-foot trees, some 6 feet in diameter, and feathered them atop one another. State officials estimated that 4 billion board feet of timber was knocked over that day—enough to build at least 150,000 homes. Some 7000 elk, deer, and bear were killed, along with millions of salmon fingerlings. Fish in rivers near the volcano literally jumped onto shore as the water temperature rocketed to 90 degrees. On the whole, the eruption turned 230 square miles of lush forest that lured

The eruption turned 230 square miles of lush forest into a lunar wasteland.

500,000 visitors a year for wilderness recreation into a lunar wasteland.

The lateral blast uncorked Mount St. Helens, opening a vent from which, for the next nine hours, a 3-mile-wide gusher of ash and magma spewed skyward. It reached 63,000 feet into the stratosphere, twice the height at which most commercial airliners fly. A group of climbers near the summit of Mount Adams, 30 miles to the west, reported that ash, pebbles, and singed pine cones rained down on them. The force of the eruption also sucked electric charges upwards from the earth's surface, turning the plume into a volatile lightning storm and igniting a 3,000-acre forest fire. Before they hurried down the mountain, the Mount Adams climbing party reported sparks jumping between their climbing axes.

The ash traveled east at 60 mph, and by 9:45 AM had turned day into night in Yakima, Washington, dumping 600,000 tons of the fine powder on the city. Two hours later it did the same thing in Spokane. Six thousand miles of Washington's highways, city streets, and logging roads were, for a time, impassable; no traffic got in or out of the town of Ritzville for three days. The fine ash clogged air filters and brought cars already on the road sputtering to a halt, stranding thousands of drivers. Regional air traffic was grounded. Up to 6 inches of ash rained down on states as far away as Oklahoma.

Even as the main eruption subsided, a series of minidisasters ensued. Water and mud began to seep out of the initial landslide and rumble down the North Fork of the Toutle River. "You could see it upstream," said Venus Dergan, who had been camping with her future husband, Ronald Reitan, on the riverbank 38 miles from Mount St. Helens. "It started snapping down the trees like they were toothpicks. The mud flow shot through the trees, and it was upon us. There was no place to run." Dergan and Reitan were swept downstream, but managed to hopscotch across the debris and back to shore.

The mud raised the Toutle 21 feet above flood stage, destroying two hundred homes, then swept into the Cowlitz River and finally into the

A car sits buried in four feet of volcanic ash at Coldwater Camp, 6.5 miles north of the summit. The ash traveled at 60 mph, leaving nearly two hundred miles of highway in Washington State heavily damaged from the eruption. Some 17 miles from the volcano, the blast flattened thousands of trees, many of them 100 feet high and 6 feet in diameter.

ANATOMY OF AN VOLCANO

Volcanoes are different than mountains in that they are formed by volcanic eruptions caused by extreme heat (known as igneous, from the Latin meaning, "of fire") released from magma—a combination of minerals, dissolved gases and molten rocks. When temperatures reach approximately 900 degrees C at the earth's mantle (lithosphere and asthenosphere) rock will melt, and because it is less dense than the surrounding crust, the magma is thrust upward through volcanic vents in eruptions.

Erupting volcanoes can spew giant clouds of ash and gas into the Earth's asthenosphere, as well as producing rivers of hot lava. They can be fairly docile or extremely explosive, depending on the thickness of the magma. Slow-moving magmas (those rich in silica, such as andestite and dacite) build up great pressure, preventing gas from escaping the vents and thereby producing highly volatile eruptions. In contrast, magmas that are more fluid and low in silica, such as basalt, will often create a steaming lava flow that is less explosive.

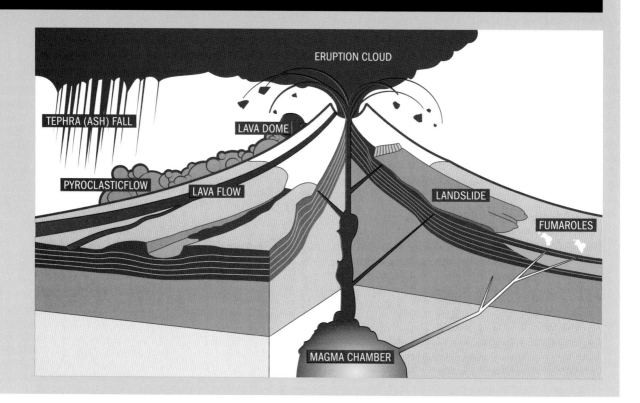

Columbia. It shrank the Columbia River's 600-foot-wide, 39-foot-deep shipping channel to just a third of its size. More than thirty oceangoing freighters were stranded upstream in Portland, Oregon, and in Vancouver, Washington.

Four days after the eruption, President Jimmy Carter flew over the disaster zone in a helicopter.

"Somebody said it looked like a moonscape, but the moon looks like a golf course compared to what is up there," the president said. "It is a horrible-looking sight. There is no way to prepare oneself for the sight we beheld this morning."

The Aftermath

Mount St. Helens caused a staggering $1 billion in damage and opened Americans' eyes to the dangers of volcanoes on their own soil—right next to their cities and homes, in fact. "There are 169 volcanoes in the U.S. that have erupted in the last ten thousand years," notes Jeff Wynn, chief scientist of the U.S. Geological Survey Volcano Hazards program. "That means they could become active again with very little warning."

After the 1980 eruption, Congress funded the U.S. Geological Survey Cascades Volcano Observatory to monitor volcanoes in the Cascades range, including Mount St. Helens, Mount Hood, and Mount Rainier. In subsequent years, observatories were established for the Long Valley caldera in California, the Yellowstone National Park region, and Alaska. Only an observatory in Hawaii predated Mount St. Helens's eruption. The budget for monitoring volcanoes went from $2 million in 1980 to $21 million in 2000, and the U.S. Geological Survey now has more than 120 vulcanologists on staff.

In many ways, Mount St. Helens represented the dawn of earthquake science in the United States. "I'd say our knowledge has easily doubled [since then}," adds Wynn. For example, Mount St. Helens's deadly lateral blast had been an unknown facet of volcanic explosions. Since 1980, scientists have catalogued more than two hundred volcanoes around the world that have erupted in a similar way.

Civilian-use GPS technology was in its infancy in 1980, but fixed GPS monitors can measure a volcano's growth and movement in real time today. That means scientists such as Johnston don't have to risk their lives by putting themselves in close proximity to active volcanoes.

READY TO BLOW

Most vulcanologists agree that Hawaii's Mauna Loa will erupt before long, but Hawaiian volcanoes are shield volcanoes, which means that spectators could pack a lunch and watch. The following are four of the planet's scariest stratovolcanoes, which are the very explosive kind capable of turning entire cities into graveyards.

MOUNT VESUVIUS, ITALY

Yes, *that* Mount Vesuvius, the volcano that buried Pompeii in AD 79, is rumbling.
WHY WORRY: More than one million people live in nearby Naples.

COTOPAXI, ECUADOR

This 19,388-foot-high monster is restless—and capped with glaciers.
WHY WORRY: Volcanic heat can turn glacial ice into muddy rivers (called lahars) with the consistency of wet cement.

MOUNT MERAPI, JAVA, INDONESIA

Indonesia has more volcanoes than any other country, and Merapi is the most active.
WHY WORRY: Locals consider the peak sacred—and fifty thousand of them live on its slopes.

GALERAS, COLOMBIA

An eruption from this notorious smokestack in 1993 killed nine people on a scientific expedition.
WHY WORRY: It has erupted nine times in 2009.

MOUNT PINATUBO

Catastrophe in the Philippines:
June 15, 1991

As calamitous as the eruption of Mount St. Helens was, an event that occurred eleven years later on June 15, 1991 made it look like a firecracker by comparison: the eruption of Mount Pinatubo on the island of Luzon in the Philippines. More than eight hundred people perished, many in poorly made houses that collapsed under the weight of ash from the eruption. Pinatubo's effects were felt not only throughout the Philippines, where thousands of acres of crops and heads of livestock were destroyed, but globally.

Pinatubo's immediate effects were devastating. Philippine skies darkened and ash rained down so heavily that hundreds of people were killed in building collapses. Drivers couldn't see without their windshield wipers, and jet aircrafts lost their engines when they ingested clouds of abrasive dust.

From an environmental point of view, the blast injected 15 to 20 megatons of sulfur dioxide gas into the stratosphere, more than 14 miles above the earth. Mixed with oxygen and water from the air, the gas solidified into 30 to 40 megatons of sulfuric acid particles. Measuring less than 1 micron across, these particles are so tiny that they stay aloft for three to four years. Within twenty-one days, high-speed stratospheric winds had spread a thin layer of the reflective particles into a ring running clear around the equator.

The Aftermath

That sulfur dioxide proliferation remained in the stratosphere for roughly three years, and as a result of its presence, global temperatures cooled by about one degree Fahrenheit during that time. Pinatubo also injected more aerosols into the atmosphere than any volcanic event since the eruption of Krakatoa in 1883. Those chemicals, along with those from the eruption of Mount Hudson in Chile two months later, greatly hastened the depletion of the ozone layer worldwide.

Technology for better monitoring of volcanoes has advanced dramatically in the last thirty years. In 1980, relatively crude seismometers could only detect high-frequency noises, such as rocks breaking. Today, broadband seismometers the size of a tin of coffee can hear the low-frequency rumblings of fluid moving far below the earth's surface. Scientists also now employ geochemistry in their predictive efforts, using planes to sniff indicator gases, such as hydrogen fluoride, above restive volcanoes.

But although we may be getting better at predicting volcanic eruptions, as long as the earth's blood boils it will sometimes storm to the surface, and when the gods of old charge down from their summits, it's best just to get out of the way.

Filipino farmers plow their rice fields in San Fernando as nearby Mount Pinatubo erupts with smoke and volcanic ash. The eruption sent a cloud of ash some 20 miles into the air, bringing complete darkness to much of the central Luzon area. Scientists estimate the eruption was ten times larger than the 1980 eruption of Mt. St. Helens.

"This is a real hazard."

EYJAFJALLAJÖKULL, ICELAND

Crippling a Beleaguered Airline Industry: March and April 2010

On March 20, 2010, Eyjafjallajökull, a small glacier in Iceland whose icecap covered an active volcano, began to spew small, inconsequential bursts of ash and vapor explosions. Three weeks later, however, the volcano erupted in earnest, sending smoke and ash tens of thousands of feet into the air. In just days, so much ash had drifted across the skies over Europe that air travel over much of the continent had to shut down completely—the worst disruption in European commerce since World War II. Estimates of industry losses ran to more than $2 billion, threatening some airlines with bankruptcy.

How was it possible for a small volcano in Iceland to cripple the airline industry and disrupt travel worldwide for more than a week?

To answer that question, one only needs to examine the events of December 15, 1989, when KLM Flight 867 managed to avoid what could have been one of the worst aviation disasters in history. Just 100 miles upwind from Anchorage International Airport, Alaska's Mount Redoubt volcano erupted, sending an ash cloud into the skies, directly in the path of the KLM flight.

Within 60 seconds, a maelstrom of microscopic volcanic glass shards shut down all four of the 747's engines. With 245 passengers on board, the plane plummeted 13,000 feet before its pilots managed to restart the engines and steer the crippled craft to an emergency landing in Anchorage. A similar event occurred in 1982 when a British Airways flight over the Indian Ocean ran into an ash cloud on a dark evening, causing all four engines to fail. The jet glided for nearly fifteen minutes before the pilot was able to regain power and land the plane safely in Jakarta.

"This is not a hypothetical hazard" says Marianne Guffanti, head of the Volcano Hazards Program at the U.S. Geological Survey in Reston, Virginia. "There have been over one hundred encounters of aircraft with ash clouds since the early 1970s, and the majority of these encounters have involved some kind of [aircraft] damage."

The Aftermath

The danger from sudden volcanic eruptions occurs when airplanes are already on a flight path, speeding toward the ash. Seismic sensors and webcams mounted on volcanoes can pick up warning signs of an eruption days in advance. Because Iceland's volcano had been closely monitored, airline officials were able to avoid any surprise encounter with volcanic ash. While the eruption took an enormous financial toll, fatalities were thereby avoided.

In other parts of the world, however, aviation officials must rely on satellites to monitor more remote volcanoes. But relying on satellites involves a certain element of risk. Their passes over a specific area may occur hours apart—during which a crowded 747 could bumble into a newly belched ash plume. Cloudy weather can obstruct the view of satellites, and as ash plumes absorb water vapor they become difficult to spot.

With improvements in technology, the airline industry is able to create a larger margin of safety against ash-induced engine failures. But the headaches to travelers and the cost for such safety measures may be unavoidable.

A satellite view of the ash cloud over Iceland and the Atlantic Ocean after the 2010 eruption of Eyjafjallajökull—a stratovolcano whose eruption was presaged by a series of earthquakes earlier that month. Smoke and ash poured over the skies, disrupting air traffic for weeks to come. Industry experts estimate the airlines incurred losses of around $2 billion due to volcanic ash over Europe.

SURVIVAL TIPS: VOLCANIC ERUPTION

You live in a small community in the shadow of an inactive volcano. One afternoon, you hear an ominous rumble and the ground starts to shake. What should you do?

➜ According to FEMA, your first step should be to listen to the radio or television for emergency information. Follow the evacuation order issued by the authorities and leave immediately from the volcanic area. Bring a battery-powered radio with you so you aware of any changes to the emergency broadcasts.

➜ If you have been ordered to evacuate, your best chance of survival is to use your car to drive to a safer area. If there is heavy ash fall around you, keep your speed to 35 mph or slower to avoid clogging your engine's radiator and air intake. Stay to areas downwind of the volcano to avoid ash, and avoid river valleys and low-lying areas.

➜ To protect yourself from ash fall, wear protective garb, including respiratory masks, long-sleeved shirts and pants, and goggles or safety glasses.

➜ If your home is far enough away and you do not need to evacuate, remain indoors until the ash has settled, unless there is danger of the roof collapsing. Close all doors, windows, and ventilation in the house, including chimney vents, furnaces, fans, and air conditioners.

There are four major natural causes of wildfires—volcanic eruption, sparks from rockfalls, lightning, and spontaneous combustion. Yet many fires today are caused by human sources; discarded cigarettes, power line arcs, equipments sparks, and arson. Though technology has improved suppression of wildfires, man is often at the mercy of Mother Nature when it comes to controlling their deadly spread.

FIRE

Centuries ago in western Europe, being burned alive was a punishment reserved for what societies considered the vilest, most unrepentant criminals and sinners: heretics. Uncontrolled fire punishes indiscriminately, ravaging everything in its path until it is stopped. Putting out a blaze is no easy task, and the firefighters who choose to protect the rest of us from fire take their lives into their hands each time they get a call.

According to the U.S. Fire Administration, 3,320 people in the U.S. died in fires in 2008, and more than five times that number were injured. The worst fire in U.S. history was the Peshtigo, Wisconsin, fire of October 8, 1871. Sources estimate 1,500 people lost their lives, and between 1.2 and 1.5 million acres of forest burned. But the Great Chicago Fire occurred on the same day, relegating the Peshtigo fire to a historical footnote.

Wildfires present a unique challenge to the fire services of the United States. Between 1999 and 2009, there were almost four dozen fires that burned more than 100,000 acres each, and eight that burned 500,000 acres or more. One of the most devastating in U.S. history happened one hundred years ago.

The men who heroically fought the wildfire ripping through 3 million acres of Idaho and Montana, which came to a head on August 20, 1910, were up against a formidable enemy.

"The forests staggered, rocked, exploded, and then shriveled under the holocaust," wrote local historian Betty Goodwin Spencer. "Great red balls of fire rolled up the mountainsides. Crown fires, from 1 to 10 miles wide, streaked with yellow and purple and scarlet, raced through treetops 150 feet from the ground."

The speed of the inferno was both breathtaking and deadly. "You can't outrun wind and fire that are traveling 70 miles an hour," Spencer wrote. "You can't hide when you are entirely surrounded by red-hot color. You can't see when it's pitch black in the afternoon."

In contrast, summer that year seemed to drag on forever, slow-cooking the upper Rocky Mountains until they were as dry as a desert. Along the Bitterroot Range that divides Montana and Idaho, the temperature in April was the highest on record. May was even hotter and drier. Barely an inch of rain fell on the forests in June, and none fell in July—this in an area that receives up to 60 inches of rainfall a year on some mountains. As the hot, waterless stretch grew in length and intensity, the pine-green forests turned parched and brown.

In the arid understory, lightning strikes and sparks from the railroad found perfect tinder. By the end of July, more than ninety major fires blazed on or near Idaho's Coeur d'Alene National Forest and the Lolo National Forest in western Montana. As many as four thousand newly recruited firefighters camped in the woods, struggling to put them out. Hundreds more minor fires were simply left to burn. The situation was so desperate by August 8 that federal foresters asked for military assistance. Troops were deployed to the front lines of the fire and to towns such as Wallace, Idaho, that lay in its path. Wallace became so dry that town officials decided to ignite dynamite for 60 straight hours, hoping that the thunderous explosions would jolt rain from the sky.

It didn't work, but by August 17 things were nonetheless looking better. Swaths of land continued to burn, but the fire seemed to be contained. Then, on August 20, the forest exploded. A bizarre cold front with 75-mph winds came howling out of the west, feeding oxygen to hundreds of fires and merging them into one great inferno. Joe Halm's firefighting crew was positioned near the headwaters of the St. Joe River in Idaho when the forest around him combusted. "As if by magic, sparks were fanned to flames, which licked the trees into one great conflagration," recalled Halm, a 1909 graduate of the forestry school at Washington State College, in a 1944 history compiled by the U.S. Forest Service. "A slight wind now stirred the treetops overhead; a faint, distant roar wafted to my ears. The men heard it—a sound of heavy wind or a distant waterfall."

The forest exploded.

A man stands at the foot of the wreckage caused by a wildfire in a heavy stand of Idaho white pine, located on Little North Fork of St. Joe River in Coeur d'Alene, Idaho. The great fire that swept through the forest was so powerful, officials decided to ignite dynamite, hoping a massive explosion would jolt clouds into producing rain.

He fought the fire with his bare hands until he blacked out from smoke inhalation.

More than a third of Wallace, Idaho, was destroyed by the Great Fire of 1910, which ultimately destroyed more than 3 million acres in Idaho, Washington, and Montana. Smoke from the fire was said to be seen as far east as Watertown, New York, and as far south as Denver, Colorado. Sailors in the Pacific Ocean claimed they couldn't see stars in the sky that night.

What they heard was the wall of fire rushing headlong toward them. As the heat grew unbearable, Halm and his crew retreated into a gravelly creek. Armed with nothing but buckets of water, they fought to maintain their haven. Around them, trees crashed to the ground and firebrands whipped through the air. One man tried to sprint away to certain death; Halm reeled him back in.

"A few yards below, a great logjam, an acre or more in extent, the deposit of a cloudburst in years gone by, became a roaring furnace, a threatening hell," Halm wrote. "If the wind changed, a single blast from this inferno would wipe us out. Our drenched clothing steamed and smoked; still the men fought." Halm's crew persevered for hours

until, eventually, the fire turned and marched northward.

Elsewhere, firefighters were not so lucky. Twenty-eight men died trying to outrun the flames in a place called Setzer Creek, 6 miles outside of Avery, Idaho. Others fled into old mine shafts, where they were charred when the tunnels became blast furnaces.

"A crew of nineteen spilled off the ridge overlooking Big Creek [in the Coeur d'Alene National Forest] and sought refuge in the Dittman cabin," recounts Stephen Pyne, the event's preeminent historian and author of *Year of the Fires*. "When the roof caught fire, they ran out. The first eighteen died where they fell, in a heap along with five horses and two bears; the nineteenth twisted his ankle in crossing the threshold and collapsed to the ground, where he found a sheath of fresh air. Two days later, Peter Kinsley crawled out of a creek."

A fifty-person crew near the Middle Fork of Big Creek, led by ranger John Bell, dove facedown into a stream as the flames leapt across the tree crowns, burning the skin on the back of their necks. A falling tree crushed three of them and seven others were roasted to death after fleeing into a hole dug out by a homesteader.

But the story that would come to define the Big Blowup of 1910, becoming part of the mythology of the West and helping to cement federal firefighting policy for the following 90 years, was that of forty-year-old Edward Pulaski.

The forest ranger was leading about forty firefighters in a retreat from a wall of flames descending upon their position at Placer Creek, 10 miles southwest of Wallace. Unbeknownst to the crew, some townspeople had set a backfire—a last-ditch attempt to clear out fuel and save Wallace from the approaching blaze. As the two fires raced toward

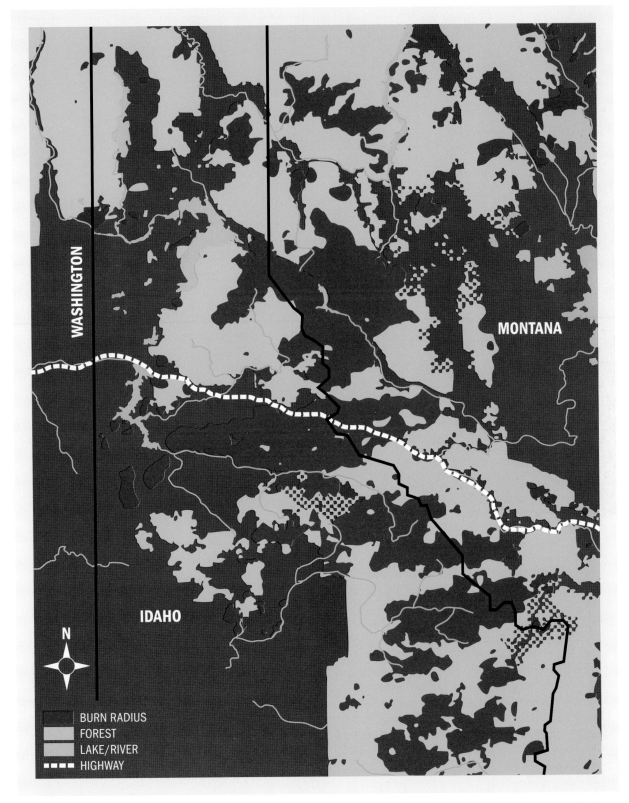

BURN RADIUS
FOREST
LAKE/RIVER
HIGHWAY

WASHINGTON

MONTANA

IDAHO

N

Fire marshals assess the damage to the Wallace, Idaho, forest area after the cold front swept across the region and released heavy rains that finally extinguished the Great Fire. Unfortunately, it wasn't in time for Grand Forks, Idaho, which was was completely incinerated, in what would become the largest fire in recorded U.S. history. The fire had cost the lives of 78 firefighters and seven civilians, and by the time it was done, the Big Blowup had ripped through 3 million acres of Idaho and Montana, traveling, at times, at speeds up to 70 mph.

them, Pulaski ordered his men into an abandoned mining tunnel and told them all to lie facedown in the mud. As heat and flames lapped at the tunnel's entrance, Pulaski covered it with blankets and fought the fire with his bare hands until he blacked out from smoke inhalation, as had the rest of his crew.

Around midnight, according to Pyne's account, one firefighter awoke and made his way to Wallace, where a search party was organized. When the rescuers reached the tunnel, five men had died—but the others survived. Pulaski was temporarily blind and his lungs were so damaged that he breathed with great difficulty, but he lived to develop the Pulaski Fire Axe, which is still widely used today.

When the sun rose on the morning of August 21, 1910, Wallace had lost a third of its town to the fire. Nearby Grand Forks, Idaho, was completely incinerated. On the other side of the range, in Montana, the towns of Taft, De Borgia, Henderson, and Haugan were all destroyed. Smoke filled the sky as far south as Denver and east as far as New York State. Sailors on the Pacific Ocean claimed they couldn't see the stars that night. Two days later, a cold front swept over the Bitterroots, releasing a steady rain, and the great fire was finally extinguished. But not before seven civilians and seventy-eight firefighters had died.

A third of the town was lost, and nearby Grand Forks, Idaho, was completely incinerated.

The Aftermath

Though the U.S. Forest Service came into existence in 1905, it was the Big Burn of 1910 that defined its mission. By the time the first flames leapt from the forest that year, the question of whether to fight wildfires was already being hotly debated across the West. Some people argued that fires are part of a forest's natural evolution. But Teddy Roosevelt conservationists, who staffed the new agency, were eager to protect forests from danger—and fire, they believed, was as perilous as clear-cutting.

The utter destruction caused by the fires of 1910, along with the heroic stand of Edward Pulaski, helped cement an antifire ideology in the Forest Service. Congress poured money into the effort and, by 1935, Gus Silcox, the head of the service—and a veteran of the Big Blowup—declared that all forest fires should be extinguished by 10 AM the following day. The service created its own army to fight fires, replete with ground troops to dig trenches and set backfires, elite smoke jumpers to parachute into remote areas, and an air force of tankers, reconnaissance planes, and helicopters.

But even as Silcox was declaring war on wildfires, some foresters and conservationists began to question whether the policy was actually healthy for the ecosystem. Fire, it soon became clear, was an integral part of forest ecology. Yet as waves of people moved into forested areas, it became even more imperative to hold fire back.

Because fire has not been allowed to thin forests naturally, land that has historically had 30 trees per acre now has between 300 and 3,000 per acre, creating plenty of fuel for the next lightning strike. In fact, the area of forestland that burns each year has tripled, from 2.5 million acres in 1994 to 7 million acres in 2002.

That aggressively fighting all fires can lead to bigger, more frequent blowups is an irony that's finally begun to be appreciated institutionally. Today, land managers in both the National Park Service and the Forest Service are at work developing fire management plans that will clarify which fires should be fought, which should be allowed to burn, and which, even, should be set intentionally.

Much has changed since 1910. Firefighters who lost their lives battling the Big Burn could have been saved with today's technology and perhaps not. And although fighting fires now is safer, it will never be completely free from danger. However, firefighters today have one weapon in their arsenal that gives them a distinct advantage: air support.

This stand of Idaho white pine, near the Little North fork of St. Joe River, Coeur d'Alene, Idaho, was devastated in the Great Fire.

It's early afternoon on October 26, 2007, on the slopes of Palomar Mountain in Cleveland National Forest, 70 miles northeast of San Diego, and California is at war. The steep hillsides are thick with fuel—which is what firefighters call the tangle of trees and scrub and grass so desiccated it crackles underfoot, as combustible as oil-soaked rags. The Palomar Observatory is next in the Poomacha fire's path.

Circling above is a cobbled-together fleet of air tankers and spotter planes in a 12-mile-wide column of airspace 1,500 feet above the fire. There are C-130 Hercules, Grumman S-2 Trackers, a giant Martin Mars flying boat, a DC-10 and, 1,000 feet above them all, in the back seat of a Vietnam-era OV-10 Bronco, the conductor of this symphony of aircraft: Scott Upton, a supervisor with the California Department of Forestry and Fire Protection's (Cal Fire) Air Tactical Group.

The tankers are arguably Cal Fire's single most effective weapon in containing Poomacha. Their mission: stop the fire's advance by laying down lines of retardant. The flying is aggressive and dangerous; since 1958, more than 130 crew members in large tankers have died. Until recently, aerial firefighting was the last vestige of a seat-of-your-pants aviation culture of tinkerers and pilots who ruled wildfire attacks for fifty years.

But the industry changed with the stroke of a pen in 2004. The U.S. Forest Service, concerned that many of the aging aircraft were unsafe,

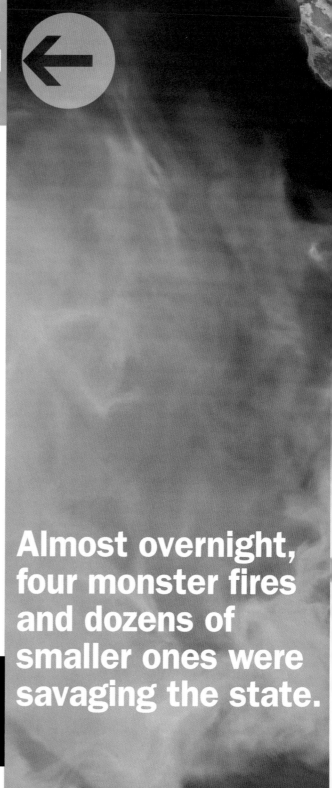

Almost overnight, four monster fires and dozens of smaller ones were savaging the state.

A satellite view of the fires in Southern California in 2007. The fires would destroy more than 1,500 homes and over 50,000 acres of land burned from Santa Barbara County to the U.S.–Mexico border. Nine people perished and dozens more were injured, as more than 6,000 firefighters from around the country worked to quell the blaze.

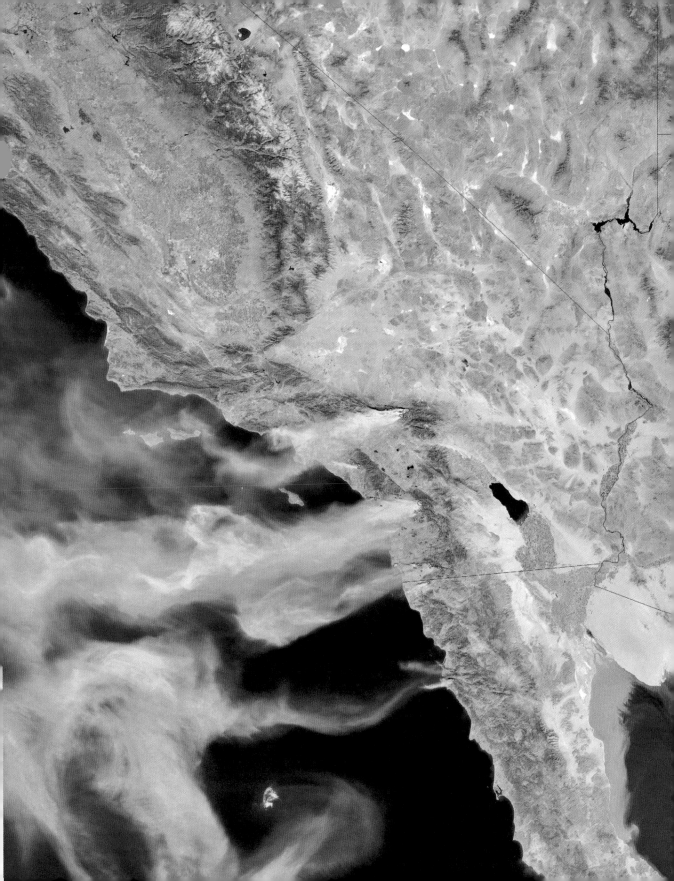

"A fire needs a stopper, and that's where massive amounts of retardant can do the job."

suspended its contracts with 75 percent of the nation's large air tanker fleet, creating a dangerous shortage. "It's very clear that there are not enough," said Rick Hatton, managing partner of 10 Tanker Air Carrier, which flies the world's only McDonnell Douglas DC-10 equipped to fight fires. "There's no one who'll argue that." As flames burned out of control in Southern California, Cal Fire and the U.S. Forest Service were forced to bring in replacement tankers from the Air National Guard, the Air Force Reserve, and even from a private firm in Canada. According to Gene Powers, whose Greybull, Wyoming–based Hawkins & Powers Aviation was a leader in aerial firefighting for decades, "California lost a lot of time."

The fires started suddenly. On Saturday, October 20, 2007, there wasn't a single fire burning in the state. Scott Upton was shuffling paperwork up in Chico, in Northern California; Tracker pilot Joe "Hoser" Satrapa was about to leave his home in Nevada City in the Sierra foothills to go elk hunting in Montana. Then, a spark here, a spark there, was fanned by the perfect storm—Santa Ana winds gusting to more than 100 mph, temperatures in the nineties, low relative humidity, and a rapidly expanding urban-wild landscape that hadn't seen measurable rain in months. Poomacha was born as a house fire on the La Jolla Indian Reservation. Almost overnight, four monster fires and dozens of smaller ones were savaging the state. Cal Fire sounded the alarm and Upton and Satrapa were scrambled to Southern California air bases in Ramona and Hemet, east and northeast of San Diego. Five days later more than 400,000 acres were charred, and the airspace over Palomar Mountain looked like Pearl Harbor.

On the hillside, hand crews were cutting fire lines. In the sky above them, working six radios in the Bronco, Upton directed the strike in conjunction with the incident commander on the ground. A twenty-year firefighting veteran with a neat mustache and steady brown eyes, Upton had spent fourteen years on ground crews; this was his first year as an Air Tactical Group supervisor. He called it the busiest job in firefighting. "I am the eyes and the ears in the air, and I see the whole picture," he said.

First, he unleashed another Bronco flying lead. It swooped down, banked east, and screamed upwind over the ridge a few hundred yards from the fire. It released a puff of smoke to mark the drop and climbed sharply. A thousand feet behind the Bronco, Satrapa pushed his Tracker's yoke forward, going to full flaps to come in at 150 feet—well below the mountains on either side of him—loaded with 1,200 gallons of reddish fluorescent fire retardant weighing more than 11,000 pounds. But Satrapa knew what he was doing: he's a former navy top gun and a veteran of 162 combat missions over Vietnam. He sighted his escape route down the hill and into a valley toward the east and pickled the load, dropping everything he had. Poof! A reddish spray shot out of the Tracker's belly and rained down on the trees and underbrush. As Satrapa flew back to Ryan Air Attack Base in Hemet to refill his tanks, one airplane after another slid in behind, building a wall of retardant.

Hot weather, a drought, and the strong Santa Ana winds, with gusts exceeding 85 mph, were major contributing factors in the extreme fire conditions. California's "fire season" traditionally runs from June to October, and wildfires are a constant threat to the many homes that are built in the state's canyons and hillsides, surrounded by dense forest.

TAKING TO THE AIR:
A PERILOUS FUTURE?

In the early 1930s, the first pilots to fight fires from the air dropped wooden beer kegs full of water from Ford Tri-Motor planes. The end of World War II ushered in the large-scale use of firefighting aircraft. Between 1940 and 1945, the United States built three hundred thousand military airplanes, many of them large bombers. When the war was over, tens of thousands of them were surplussed. The planes weren't just cheap; they were designed for carrying and dropping heavy loads in extreme flight regimes—perfect for converting to flying tankers. Gene Powers and other World War II vets welded holding tanks where bombs and torpedoes had been carried and went to work.

Tactics and strategy evolved: smaller, more maneuverable planes and helicopters were best for an initial attack—surgical water or retardant strikes of up to 1,000 gallons per load on fires that were just breaking out. The real muscle, however, was what became classified as the Type 1 tanker. A big bomber, it is capable of dropping thousands of gallons at a time in an extended attack, in effect erecting walls by reinforcing cleared lines built by hand crews on the ground to block a fire's path.

The flying, said Tracker pilot Joe Satrapa, was always "right on the edge." A few years ago he hit a sudden downdraft and clipped 10 feet of pine trees, which took out a 3-foot section of his left wing. "It's all based on experience, assessing the terrain and the wind," he said. "In one second at 140 knots you travel nearly the length of a football field, so if you're late or early by one second you miss the target. And when you pickle (drop) the load, you've got 10,000 pounds of instant lift; I've got to push the yoke all the way forward. You've got to have a warrior mentality and be ahead of the plane every time—or you'll die."

But more than a decade into the tanker era, two things happened. The myriad federal agencies that traditionally hired tankers, from the Bureau of Land Management to the Forest Service, united into a single National Interagency Fire Center, in Boise, Idaho, staffed, as Powers noted, "not by aviators, but by tree farmers." And all those old airplanes became ever older and more worn out. The tree farmers wanted more modern equipment—turbines instead of pistons—but those planes were harder to come by and much more expensive to buy and operate. As the decades passed, neither the Federal Aviation Administration nor the Forest Service policed the industry. "Oversight slipped through the cracks," said 10 Tanker Air Carrier's Rick Hatton. No one took responsibility except the operators themselves. Powers and other contractors acquired a handful of C-130s and P-3 Orions from the federal government in the 1980s, but even they had thousands of flight hours on their airframes. The National Interagency Fire Center wanted cheap prices, but also performance and safety; something had to give.

In 2002, it did: on June 17, the wings on a forty-five-year-old C-130A, owned by Hawkins & Powers and loaded with 2,700 gallons of retardant, folded in flight, killing all three crew members. A television crew recorded the accident, amplifying its impact. A month later, a fifty-seven-year-old Hawkins & Powers–owned PB4Y-2 suffered structural wing failure during a fire in Colorado, killing two. These were hardly the first tanker crashes but because one came on the heels of the other, they made headlines, and the Forest Service established a blue-ribbon panel to study the issue. Two years later, it canceled contracts with companies that operated thirty-three of the nation's forty-four large air tankers. Within months, Hawkins & Powers went out of business.

TANKER FUNCTION

	SIZE	DROPS*	PAYLOAD (GAL.)

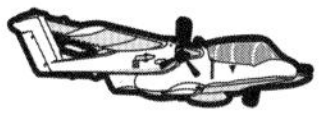

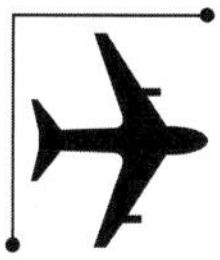

OV-10 BRONCO
41 ft.
40 ft.

Originally used for antiguerrilla operations in Vietnam, this two-seater carries the mission operator to monitor the airspace and direct tanker runs, circling 1000 ft above the other planes.

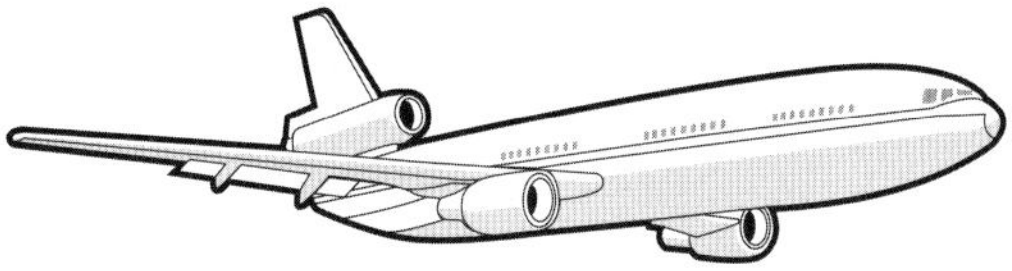

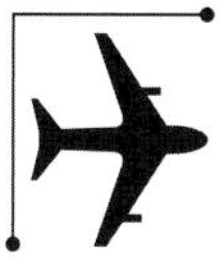

DC-10 (TANKER 9100)
182 ft.
155 ft.

1

12,000

This former passenger jet is now contracted out—at $5500 per flight hour— to lay down straight lines of fire retardant. The three-man flight crew can unleash its payload in 8 seconds.

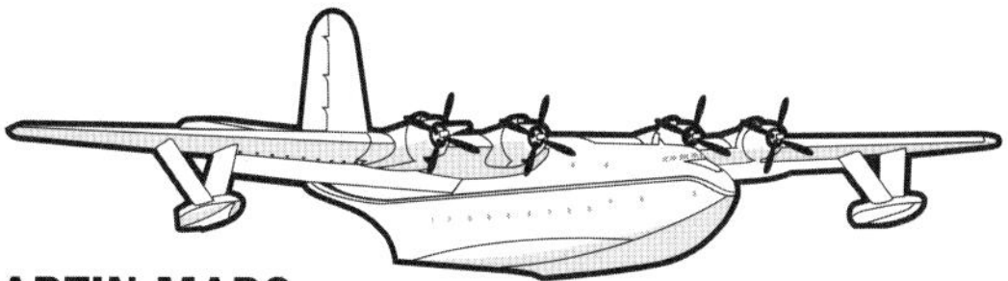

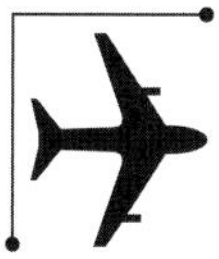

MARTIN MARS
120 ft.
200 ft.

1.7

7,200

Designed in 1938 as a Navy bomber, this flying boat has found its true calling. To fill its tanks, the Mars lowers intake scoops and skims a body of water at 70 knots.

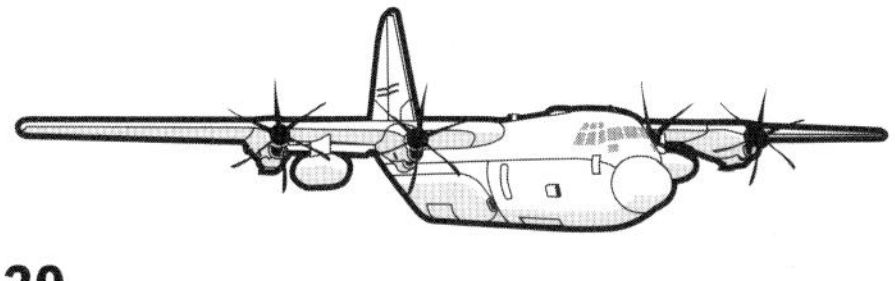

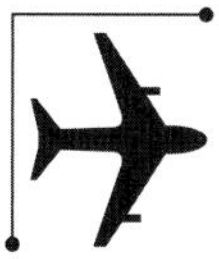

C-130
98 ft.
132 ft.

5

2,700

The military's 45-year-old transport workhorse pours 2700 gallons of retardant from a modular system in its belly. The C-130s have fought fires as far away as Indonesia.

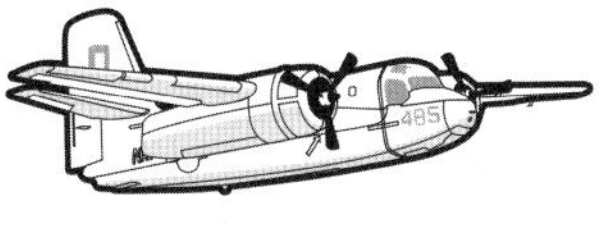

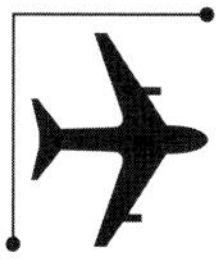

GRUMMAN S-2F TRACKER
43 ft.
72 ft.

10

1,200

The retrofitted "stoof" carries a relatively small payload—but its 230-mph top speed and precise handling make the Tracker perfect for initial attacks in tight spaces.

* The number of drops needed to equal the volume laid down by the DC-10 on a single pass.

There were several ignition sources in the California wildfires of 2007. Power lines were damaged by high Santa Ana winds and one fire was known to have started when a semi-truck overturned. Other fires began due to arson, and a 10 year-old boy later admitted that he was playing with matches and accidentally started the Buckweed Fire.

As the California fires boiled, Scott Upton and his colleagues at Cal Fire hopscotched from one air attack base to another, initially relying on eleven helicopters, twenty-three Trackers, and thirteen Broncos, all owned by the state but maintained and flown by pilots working for DynCorp, a contractor. At first the winds were too high to fly any aircraft; in crosswinds above 30 knots, the retardant—a mixture of water, ammonium sulfate, and iron oxide—disperses too much and is ineffective. Driven by the wind and the hot, dry conditions, large embers jumped fire lines and spread quickly. "All we could do," Satrapa said, "was real-time structure protection" by dumping retardant on and around threatened houses and other buildings. As the winds calmed, planes and helicopters began playing catch-up, but the Trackers' 1,200-gallon drops weren't enough. "The Tracker is effective as an initial attack tanker," Powers noted, "but bigger is better." California faced what he called an "outrageous shortage" of large tankers.

"When you build lines, you need heavy tankers," Upton said. That's because, from the ground, all you can do is chase a big fire. "You can't get in front of it and stop it," Hatton added, "because the fire might be moving at 15 or 20 mph, while ground crews with axes and bulldozers might move at 1 or 2 mph. A fire needs a stopper, and that's where massive amounts of retardant can do the job." It takes ten Tracker drops to equal the volume of one from the DC-10, for instance, and the two are still

not equally effective: "The smaller planes require more precision," Hatton said, "and the total amount of retardant is greater because you have to overlay the line on every drop. A line with a break in it is virtually useless; fire will find the break."

Under mounting pressure, California called in Hatton's DC-10, which had been certified as airworthy by the Federal Aviation Administration but still hadn't been accepted by the Forest Service. The state also recruited a rare and enormous dinosaur of an airplane: a sixty-year-old, four-engine, piston-powered Martin Mars flying boat out of Vancouver, British Columbia, which could carry 7,200 gallons of retardant. And it mustered four Air National Guard and two Air Force Reserve C-130Hs, which use a system of pallet-mounted, 2,700-gallon tanks that roll into a Herc's belly and discharge retardant out the tubes in the rear of the plane.

The DC-10 was based in California, but getting the Air National Guard and Air Force Reserve airplanes (which flew in from Colorado, Wyoming, and North Carolina) and the Martin Mars took days—a critical delay as the fires raged out of control. Two California Air National Guard C-130s tasked as air tankers never left the ground; they were new J models, and the tank pallet system didn't fit them.

As the Poomacha fire entered its fourth day, the Cal Fire base in Hemet was cranking. Every few minutes a Tracker buzzed into what's called the pit

"You hate to see fires but if they happen, I like to be the guy who puts them out."

for a quick "hot load"— getting refilled with retardant without shutting down its engines— before taking off for another drop. Each plane often made twenty drops a day. On the flight line, the noise from eleven planes and twenty-two screaming turbines was intense, the smell of distant smoke and jet fuel acrid. Above the fire, Upton was, as he put it, "calling in everything I could": the DC-10 from Victorville, California, the C-130Hs operating out of California's Channel Islands Air National Guard Station, the Martin Mars off nearby Lake Elsinore.

Flying clockwise at 2,500 feet, he gathered the tankers in a counterclockwise circle 1,000 feet below, then unleashed them behind a Bronco flying lead, the Trackers for surgical strikes. While working with the fire crews on the ground, Upton kept the airplanes separated in a dynamic and extreme flight envelope as the air assault unfolded.

The Aftermath

By the afternoon of Sunday, October 28, after five days of flames and two of heavy bombing, the Poomacha, the last of the big fires, was 95 percent contained. The Poomacha destroyed more than forty houses on the Rincon, La Jolla, and Barona reservations. As the sun dropped low, one Tracker after another returned to Hemet, followed by Upton in the Bronco. The pilots gathered under a stone pergola overlooking the flight line as a stack of pizzas arrived. Upton and Satrapa wearily plopped themselves onto lawn chairs. Both had been working for a week straight, but they were happy. "You hate to see fires," Upton said, taking a dip of tobacco, "but if they happen, I like to be the guy who puts them out."

Satrapa nodded. "We did good work today," he added. "We put in so much retardant it was glistening in the sun!"

Upton had just flown two, three-and-a-half-hour shifts. He called in the Martin Mars twice, the DC-10 twice, six C-130Hs, a P-2 and a P-3, two DC-7s, and every Tracker. But with fires, there is no telling what the night might bring, unless the skies unleash some rain. "All you can do is fight until the weather changes," he said. "You build solid lines around the flank, cut the head, and control the rear. I used every airplane I safely could. We laid over 150,000 gallons of retardant out there today."

THE 10 MOST DEVASTATING U.S. WILDFIRES

Location	Year	Acres Burned
Alaskan Wildfire	2004	6.6 million
Peshtigo Fire (Wisconsin, Michigan)	1871	3.7 million
Great Fire (Northwest U.S.)	1910	3 million
California Wildfires	2008	1.56 million
Yacolt Burn (Eastern Washington)	1902	1 million
Yellowstone Fires	1988	793,880
Murphy Complex Fire (Idaho)	2007	653,100
Adirondack Fire	1903	637,000
California Wildfires	2007	500,000
Great Hinkley Fire (Minnesota)	1894	300,000

SURVIVAL TIPS: FIRE

To protect your family in the event of a house fire, install smoke alarms in your house, practice fire drills, and plan an escape route in advance. But what happens if you're in a hotel, or at the bank or the grocery store, and the building begins to burn? Here are four tips that may save your life.

➔ Look for emergency exits as soon as you enter an unfamiliar area.

➔ Most people die of asphyxiation in a fire, not burns. To avoid being overcome by smoke, get down low and crawl; smoke rises.

➔ If your clothes catch fire, do what you were told in elementary school: stop, drop, and roll. Running increases the airflow to the fire and will make it worse.

➔ Burns come in three degrees: first (the mildest), second, and third. Immediately cool a first-degree burn by holding it under cold running water for three to five minutes. Do not put butter or lotion on the burn. For second- and third-degree burns, seek medical attention right away.

SNOW AND ICE

Wintertime is an enchanting mix of beauty and treachery. Before the modern era, winter was a time for people to huddle together against the cold and the darkness and hope for survival. Advances in science and technology have reduced the dangers of winter for the majority of the world's population. But those achievements can sometimes lull us into a false sense of security. With increasingly accurate weather prediction, practical measures such as salt distribution and snow removal, and modern conveniences like windshield defoggers and heated roads, the likelihood of a deadly winter event sneaking up on us is ever decreasing. Yet when the effects of winter take us by surprise, the results can be devastating.

← Most avalanches do not claim any victims because of the remote locations at which they occur. But there are notable exceptions. The Wellington avalanche killed nearly 100 people in Washington State in March 1910, and it's estimated that more than 50,000 soldiers were killed during World War I during the mountain campaign in the Alps at the Austrian-Italian front.

THE SUPERSTORM OF 1993

Mother Nature Rages Across the Eastern United States: March 12, 1993

Some called it a superstorm; others, the storm of the century. Both names were well-earned. When three independent weather patterns converged in the Gulf of Mexico on March 12, 1993, the resulting storm system caused three days of crippling snow, whirling seas, coastal flooding, blizzards, tornadoes, and bone-chilling cold. The amount of snow and rain that fell during the storm was almost biblical—44 million acre-feet—"comparable to forty days' flow on the Mississippi River at New Orleans," according to a National Climatic Data Center report.

The Superstorm of 1993 was a freak of nature. Its seeds had been sown a month earlier as the jet stream—the river of air that flows east to west 30,000 feet above North America—dipped dramatically down from Canada and blasted cold arctic air across Montana, Wyoming, Colorado, Kansas, and Texas, and into the Gulf of Mexico before banking back up the eastern seaboard. The jet stream that spring looked like a giant *U* flowing over the central and eastern United States.

On Friday, March 12, a cluster of powerful thunderstorms formed in the northwestern Gulf of Mexico and then merged with a narrow band of snow and rain that was pushing in from the West Coast. The two storm systems collided with the jet stream, and by 7 PM they began screaming across the Gulf toward Florida, pushing a large storm surge ahead of their path.

"The sea conditions were absolutely incredible," Coast Guard Petty Officer Rob Wyman told the *Washington Post*. "It looked like a big washing machine. There were huge waves and spray and hail." The sea was so powerful that the 200-foot freighter *Fantastico* sank 70 miles off Fort Myers, Florida. Three people were rescued from the 50-mph winds and 30-foot swells, but seven crew members died when Coast Guard helicopters ran low on fuel and had to return to base. "They were just getting the hell beat out of them," Wyman said. "I felt for them. They were just holding on, either too weak or too scared to grab the basket.... We were so close, so close to getting these guys, and we did absolutely everything we could. And yet we couldn't get them."

Ten miles from Key West, the 147-foot freighter *Miss Beholding*, hauling M&Ms and candy bars, ran aground on a coral reef. Several charter fishing vessels and sailboats sank in deeper seas. By the time the superstorm had passed, the Coast Guard had deployed more than one hundred planes, helicopters, and boats, rescuing 235 people from more than one hundred boats in the Gulf of Mexico.

But the storm was far from done causing devastation. It blew ashore, loaded with the classic ingredients for tornado creation: cold, dry air colliding with a warm, moist front. In Florida, the storm spun off no less than fifteen twisters, which overturned mobile homes and launched trees and other debris like missiles through the air. Between 4 and 5:30 AM on Saturday, a storm surge as high as 12 feet in some places swept ashore and drowned at least seven people along Florida's west coast.

Temperatures on land dropped below freezing, and 6 inches of snow fell on the state's normally balmy panhandle. Further inland, the temperature in Orlando fell to 33 degrees—9 degrees below its record low for that day. The storm left forty-four Florida residents dead in its wake.

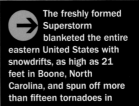

The freshly formed Superstorm blanketed the entire eastern United States with snowdrifts, as high as 21 feet in Boone, North Carolina, and spun off more than fifteen tornadoes in Florida. Millions were left without electrical power, and the storm caused more than $6 billion in damages.

"It looked like a big washing machine."

10 MOST DEVASTATING SNOWSTORMS IN THE U.S.

Name	Location	Deaths	Characteristics
Schoolchildren's Blizzard of 1888	Nebraska	500	-40 degree temperatures
Blizzard of 1888	New England	400	40-50 in. snowfall
Storm of the Century 1993	Eastern U.S.	270	60 in. snowfall in Tennessee
Great Lakes Storm of 1913	Great Lakes Region	235	12 ships sunk
Armistice Day Blizzard of 1940	Northern Midwest	154	Winds of 80 mph
Knickerbocker Storm of 1922	Eastern Seaboard	98	22,000 Sq. Miles Affected
Halloween Blizzard of 1991	Northern Midwest	22	37 in. snowfall in Minnesota
April Fools' Day Blizzard of 1997	East Coast	n/a	33 in. snowfall
Great Blizzard of 1899 (The Snow King)	Eastern U.S.	n/a	-35 degrees in Virginia
Blizzard of 1966	Rochester, NY	n/a	103 inches of snow

By 7 AM the storm had reached Atlanta, dropping massive amounts of snow. Officials closed the airport, stranding three thousand people, and residents used snowmobiles and skis to get around.

At the Birmingham, Alabama, office of the National Weather Service, meteorologists couldn't quite believe what they were seeing on their computers. "The models must be nuts," meteorologist Brian Peters said on March 12, as the storm barreled toward his city. "We're looking at 12 to 18 inches of snow…. There's just no way. That's 50 percent higher than any other previous record." The models proved right. Birmingham got 13 inches of snow, and temperatures dropped to a record 2 degrees above zero.

In Boone, North Carolina, snowdrifts stacked 21 feet high. Semi trucks began jackknifing as the interstates iced up, causing miles-long delays and fatal accidents. Interstate 75 experienced an 80-mile delay that stretched from outside Atlanta all the way to Tennessee. In the backcountry, hundreds of hikers had to be rescued along the Appalachian Trail.

In the mountainous region of Georgia's Walker County, eight people in two jeeps were stranded when the storm set in. "It started snowing so hard you couldn't see," Karen Thompson told the *Atlanta Journal-Constitution*. "There was lightning all around, and then trees started cracking and popping and falling in front of us. The men would get out and chop, but the trees kept falling." Seven members of the group were rescued after being spotted by a Georgia State Police helicopter on Sunday morning; one had died while hiking out for help.

And the storm just kept going: 4.5 feet of snow fell in Mount Le Conte, Tennessee; in Latrobe, Pennsylvania, it drifted 10 feet high. For the first time in history, every major airport along the eastern seaboard shut down.

By late Saturday afternoon, the storm had descended on the major cities of the Northeast. Television coverage was nonstop. The nation's capital looked completely abandoned as people hunkered down in their homes.

The first of 12 inches of snow began falling on New York City by 7 PM; 1,700 plows methodically

"Trees started cracking and popping and falling in front of us."

cleared the streets, creating thick canyons of snow and ice that wouldn't melt for days. Snow in parts of New Jersey was capped by a dangerous layer of sleet.

As the storm continued to push north, it washed eighteen vacation homes into the sea on the eastern end of Long Island. Cities like Syracuse, New York, received almost 3 feet that day. Across New York State, twenty-three people died: like those in neighboring New England, most of the deaths were due to traffic accidents and heart attacks brought on by shoveling snow.

The single most deadly event of the superstorm would be its last. On Sunday, March 14, the 586-foot freighter *Gold Bond Conveyor* was hauling 24,000 tons of gypsum through the North Atlantic 200 miles south of Nova Scotia and parallel to southern Maine—and heading straight into the storm. The captain reported that 90-mph winds and 100-foot waves were battering the ship, and it was beginning to list.

Canada's military dispatched a hulking, four-engine Aurora turboprop patrol plane to keep watch over the *Gold Bond*. Around midnight, the *Gold Bond*'s captain, Man Hoi Chan, radioed that the ship was listing 20 degrees; the Aurora began circling just 150 feet above the water.

"He just got hit by a huge swell, and he went down," said Wongkee, who watched through an infrared camera as the ship rolled. "You could hear him telling everybody to abandon ship just after we'd flown over him. We'd gone out a couple of miles, turned around, and come back over the top where the ship had been, and everything was gone." The entire crew—twenty-nine people from Hong Kong, three from China, and one from Taiwan—died in the frigid waters. Just one body was recovered when Canadian rescue helicopters arrived. Months later, another body washed ashore on the coast of Ireland, putting a sober close to the storm of the century.

The Aftermath

Up and down the eastern seaboard, the superstorm caused up to $6 billion in damage to cars, homes, and businesses. It left 2.5 million people without power and killed 318. But thanks to early warnings—and a light-traffic weekend—fatalities were most likely lower than they could have been.

Florida received the brunt of the storm: a federal disaster was declared in twenty-one of its sixty-one counties. In all, eighteen thousand homes were damaged—particularly painful because the state was still recovering from 1992's Hurricane Andrew. There was a round of finger-pointing among Florida governor Lawton Chiles, local emergency officials, and the National Weather Service. The governor wanted to know why people hadn't been evacuated before the deadly storm surge hit; emergency officials said the National Weather Service, run by the National Oceanic and Atmospheric Administration (NOAA), didn't notify them of the problem.

While the National Weather Service had warned of a "storm of historic magnitude," it acknowledged that local officials didn't always get the message—or appreciate the gravity of the situation. After the storm, the NOAA initiated a thorough overhaul of how threats were communicated. Perhaps more importantly, the National Weather Service realized it had a weakness in predicting specific local threats. It accelerated development of numerical prediction models for everything from Gulf-based storms to local storm surges and snowfall predictions. As a result, the National Weather Service increased its snowfall-prediction accuracy rate to 75 percent—up from just 37 percent at the time of the superstorm.

AVALANCHE OF HELENA, MONTANA

**Serene to Surreal in Seconds:
February 17, 2007**

Some people would like nothing more than to pull on warm pajamas, get a fire going, and hibernate, bearlike, until winter is over. Others, like skiers, snowboarders, and snowmobilers, treat snowy mountains as a playground, spending as much time outdoors as they can before the spring thaw. But winter sports can be as deadly as they are exciting, as three snowmobilers found out on February 17, 2007.

It was the weekend they'd been waiting for all year. After an early season marked by disappointing snowfall, friends Jason Crawford, 27, Brett Toney, 27, and Kris Rains, 26, awoke that Saturday to find the hills around Helena, Montana, buried in fresh powder.

"It was the best snow we'd had all season," Crawford said. "We were pretty anxious to get out there."

The snowmobilers' destination was a remote spot called the North Fork Bowl, deep within the Big Belt Mountains. They had gone high marking in the surrounding range—riding their snowmobiles up steep slopes and turning out of the climb at the last possible moment before stalling or tipping— yet they had only managed to reach this spot once before. From the trailhead it took an hour of hard technical riding, but when they arrived, the sight of the bowl took their breath away—a wide-open amphitheater covered in chest-high snow: sled-head paradise. As usual, Toney gunned his engine first, racing straight up the 40-degree slope, with

The most powerful avalanches have the capacity to mix ice, rocks, trees, and other mountainous materials on the slope. Powder snow avalanches are the largest and most powerful, capable of reaching speeds up to 170 mph. The majority of fatalities are the result of dry slope avalanches, when the densely packed snow on top begins to slip.

"My first thought was, 'I'm going to die.'"

Crawford close behind. At the top of their run, both machines bogged down in the deep, soft snow. Climbing off his sled, Crawford looked at Toney. His friend stood about 30 feet upslope and the same distance to his right. Suddenly, Toney bolted, leaping across the snow before falling on his stomach.

"What the…?" Crawford said. Then he turned up the hill and saw it: a wall of white 4 feet high and 300 feet across thundering down on them. "I turned, took two big lunges, and fell on my stomach," he recalls. "My first thought was, I'm going to die."

Eight hundred tons of snow slammed into him like a speeding freight train. Buried in the roaring tumult, Crawford tried to struggle, but the heavy snow was so thick he could barely move his arms and legs. "Every once in a while," he said, "I'd see a flash of light."

Then everything went dark.

The horror that struck the Montana snow-mobilers was not an isolated incident. Throughout the 1950s and 1960s, avalanches caused more than fifteen deaths per year in the United States. But in the 1970s, as more people began venturing into the backcountry on skis, snowshoes, and snowmobiles, the number of fatalities started creeping up, reaching highs of thirty or more per season. Since 2002, avalanches have killed an average of twenty-five people in the United States each year. And they are rarely flukes. Ninety-two percent of the victims are caught in slides triggered by themselves or their companions.

Snowmobilers can be particularly at risk. "Machines are powerful enough now—and the riders good enough—that they can get onto avalanche-prone slopes that they could never have reached ten or fifteen years ago," says Doug Chabot, director of the Gallatin National Forest

Avalanche Center in Bozeman, Montana. "And in any given day, a snowmobiler is going to hit a lot of slopes, all over the place." It's like playing Russian roulette and pulling the trigger over and over. Although snowmobilers make up about half of all avalanche deaths, snowshoers and skiers are also at risk, since one person's weight can be enough to trigger a slide.

Avalanches may be deadly, but they're also spectacular. The majority of fatalities are the result of dry slab avalanches, the kind that caught Crawford and his friends. Temperature throughout the snowpack can vary greatly, from around 32 degrees near the ground to 5 degrees or colder at the top layer. A steep temperature gradient of 6 degrees per foot can result in depth hoar, a faceted layer of crystals near the ground that fractures easily and can cause the stronger, densely bonded snowpack above it to slip. The snow Jason Crawford was riding on had fallen atop a similar type of faceted layer, called surface hoar. "These are beautiful, feathery, big flakes," Chabot says, "and they're really weak. They just won't hold a whole lot of weight above them before they all tip over, literally, like dominoes."

As the friends high marked on the fresh powder, this layer of tiny ice dominoes collapsed in an outward-expanding wave that went ripping up the hill. When it reached the steepest part of the slope, the snowpack fractured in a long, horizontal band called a crown. At that instant, everything beneath it slipped away en masse.

Up to then, the process had been practically invisible. Although very little about an avalanche can be discerned by the naked eye until snow comes roaring down the mountainside, there are some known risk factors. Eighty percent of avalanches occur during or within forty-eight hours of a snowstorm, which adds weight on top of

fragile layers of snow. Another factor is strong winds, which can deposit snow into dangerously high drifts. A release will usually occur on slopes of 25 to 50 degrees that are above the timberline and face away from prevailing winds. An abrupt rise in temperature can also weaken the structure of the pack as the snow melts.

About 80 percent of avalanches occur during or within two days of a snowstorm, which usually adds weight to already fragile layers of ice and snow packed onto a mountain. Strong winds add to the instability, creating high drifts of snow. Then, it's only a matter of the right temperatures, slope, and conditions. The result? Hundreds of tons of snow descending down a mountain.

Researchers in the United States, Canada, and Europe are struggling to better understand snowpack dynamics in order to forecast avalanches more accurately and help the public avoid and survive them. Yet even the most seasoned expert can only assess the snow's condition accurately by digging a 5- to 6-foot snow pit and smoothing the wall to get a clean look at a cross section of snow. The procedure takes a long time and is only applicable to a particular stretch of snow. Farther up a hill, or on the other side of a bowl, might be a completely different story.

To speed the process of analyzing the snowpack, researchers are developing an instrument called a penetrometer. When pushed into the snow, it can generate an instant profile of the snow layers on a handheld computer. "You can get many observations in the same time it would take you to do a single manual profile," says Bruce Jamieson, director of the Applied Snow and Avalanche Research program at the University of Calgary in Alberta, Canada.

Of course, all the information and understanding in the world isn't going to do any good if it isn't

Avalanches can accelerate 10 million tons of snow to 200 mph in seconds.

heeded by the folks out in the backcountry. The unfortunate fact is that too many people fail to treat avalanches with respect. "We all knew that it was bad avalanche conditions," Jason Crawford admitted. "Brett had seen the bowl slide before. But we kind of thought we were invincible, you know?"

That kind of attitude is all too common. Everyone who loves the backcountry knows what it's like to get powder fever—the urge to go play in the snow and be the first to lay down tracks. Avalanche researchers understand the compulsion. So a major focus is figuring out how to prevent bad decision making. After fifty-eight people died in North American slides during the disastrous winter of 2002–2003, the Canadian Avalanche Association developed the Avaluator, a wallet-size card to help users objectively assess the threat level. On one side is a checklist of simple questions about possible dangers: "Are there signs of slab avalanche activity?" "Are there signs of unstable snow?" On the other side, a chart cross-indexes the type of terrain with the updated avalanche forecast danger level. In each case, the card yields a go/caution/no-go advisory.

"What these cards do is, they jog our memory," Chabot says. "They make us think twice."

Ultimately, the goal of avalanche research is to save lives. And as it happens, the past few years have actually seen a modest decrease in the fatality rate. But can science take the credit? Most experts suspect not. "The condition of the snowpack is probably the biggest factor," Jamieson concedes. "In the two previous winters, we generally had better snowpack, with lower levels of avalanche danger." Odds are, that lucky streak won't last. With more and more people hitting the backcountry, fatalities could easily rise again.

In Jason Crawford's case, his luck hadn't totally run out. As the avalanche slowed to a stop, he came to rest with his head above the surface. Spitting snow out of his mouth, he wheezed for air and slowly struggled to free himself. His helmet had been ripped off, and one of his ankles was badly sprained. He called for his friends and tried to dig, but found no sign of them. The avalanche debris—massive, icy blocks of snow—was 10 to 15 feet deep.

"Yelling for these guys, knowing that my best friends are under the snow suffocating and there's nothing I could do about it--that's the worst feeling in the world," he recalls. "I wouldn't wish that on anybody. I'll remember it forever."

Ahead lay an ordeal almost as dangerous as the avalanche: digging out his snowmobile, and fighting his way back through miles of punishing, technical riding with a damaged machine. Twice he became stuck and only managed to free himself after hours of work. Finally, his machine died for the last time—just after delivering him to a well-traveled logging road. He hobbled another 2 miles on his bad ankle before he managed to find help. Not until the next day was a search team able to recover the bodies of Kris Rains and Brett Toney.

The Aftermath

That day changed Crawford's life. For a month he was in a wheelchair, then on crutches. A year later, he still hadn't made a full physical recovery. Emotionally, the wounds were deeper. He was determined to get back on his snowmobile and go high marking, but with a different outlook based on experience: he would stick to well-traveled terrain that he knew would be less likely to slide.

"We talked a lot about avalanches," he said. "We just never thought it could happen to us. I don't know a better way to put it to make it sink in for somebody else: it can happen to anybody."

ANATOMY OF AN AVALANCHE

Each winter outdoor enthusiasts, especially snowmobilers, venture into the backcountry, lured by the solitude and untracked expanses. There, far from any ski patrol, one of the biggest risks they face is a slab avalanche, in which a strong layer of snow breaks from the slope above it like a pane of glass and roars down the mountain with churning, deadly force. Here's how it happens:

00:00

A snowmobiler prepares to high mark a backcountry mountain slope. The open expanse may show signs of wind-deposited or eroded snow and previous avalanche activity, like fracture lines, making it a prime location for a slide.

00:22

When the cracks in the weak layer reach the steepest part of the slope, the strong layer fractures—typically with the snowmobiler sitting squarely in the middle of the slab that's about to slide down the mountain. Only then does the rider realize that something is wrong.

00:24

The fracture boundaries—crown line above, staunchwall below, and the flank walls—form and the slab breaks off almost instantly, moving downhill in a body even as it starts to disintegrate. Slabs can be as thin as a few inches and as thick as 20 feet. Unable to maintain control of the snowmobile, the rider is tossed into the snow.

00:10

Avalanches occur mostly on 25- or 50-degree slopes. Between the ground layer of the snowpack and the surface, a steep temperature gradient of 6 degrees F per foot causes depth hoar, a weak layer of large faceted crystals that doesn't bond well. This layer lies beneath a stronger layer made of rounded, densely bonded crystals.

00:20

As the snowmobiler heads upslope, his weight triggers failure of the depth hoar. Tiny cracks propagate outward—up, down, and even around corners at 300 to 400 mph. Skiers often hear a "whumping" sound as the weak layer collapse, but snowmobiles drown out this warning, so the rider continues on his path.

00:26

The slab moves past the staunchwall and flows downhill; the snow, roiling like a river's rapids, sucks the rider in. Disoriented and fighting for air, he tumbles in the snow, which can reach speeds of up to 200 mph in seconds. A large slide can release 300,000 cubic yards of snow, equaling 20 football fields filled with snow 10 feet deep.

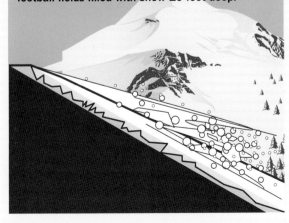

00:32

The slide fans out as the terrain flattens. The pulverized snow, which has a density of up to 40 percent, sets into a heavy, chunky mass within seconds. The rider, almost comp   letely buried in the avalanche's deposition zone, will find it hard even to move his fingers. If he's not injured, digging himself out could still take hours.

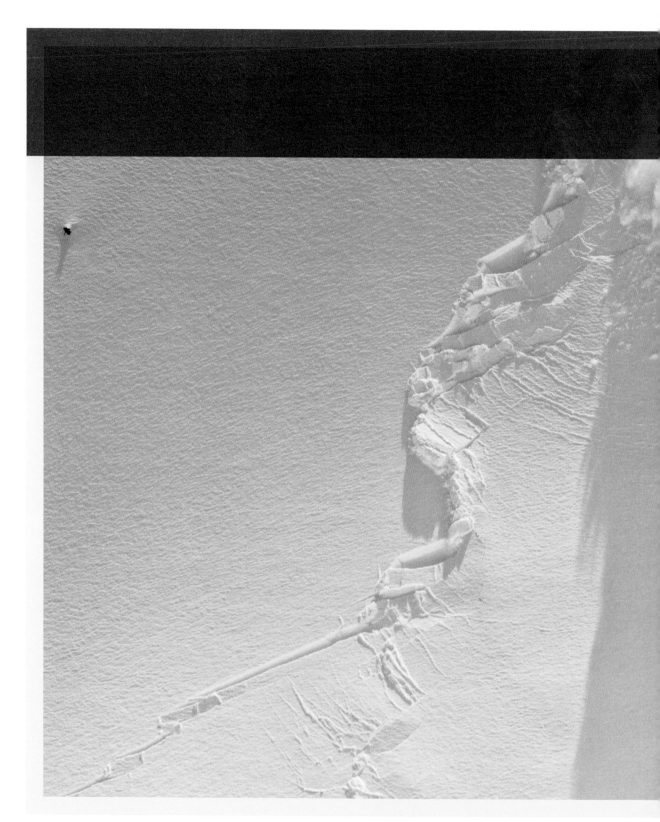

SURVIVAL TIPS: AVALANCHE

When skiing in the backcountry, a low rumbling from above is one of the most horrific noises imaginable. A skier has only seconds to react. Anyone who seeks out high-altitude locations for the fresh powder needs to be prepared.

→ Never go it alone. If you're out solo and you get buried with no one to dig you out, you'll almost surely die.

→ The survival rate among victims buried less than fifteen minutes is 90 percent. By thirty minutes, that rate drops below 50 percent, and after two hours, it's virtually nil. Seventy-five percent of victims die from asphyxiation; trauma from hitting rocks and trees makes up most of the remainder of the fatalities.

→ If caught in an avalanche, let go of everything you're holding and try to swim to the top. When the slide begins to slow, use your hands to create a pocket of air around your face, and thrust an arm or a leg toward the surface.

→ A slope that's steep enough to ski is steep enough to slide. When crossing a potential avalanche path, travel one person at a time. Evidence of recent slides or heavy snow are major red flags.

→ Each group member should have an avalanche transceiver, which can help rescuers find people who get buried; a probe, which can pinpoint their locations; and a shovel to dig out buried companions.

TORNADOES

Tornado. Twister. Cyclone. Waterspout. Dust devil. Funnel cloud. All are names for the same deadly phenomenon. Tornadoes form from thunderstorms and are fairly brief, spectacular events, typically appearing out of nowhere, tearing around, and then disappearing. Within minutes, the only evidence of a tornado's passing is the destruction it leaves in its wake.

In 1925, the word *tornado* wasn't even in the vocabulary of the U.S. Weather Bureau (now the National Weather Service). The word had been banned since 1887, when the U.S. Army Signal Corps managed the country's weather forecasting. Tornadoes were utterly unpredictable, the logic went, and forecasting them, besides being a fruitless venture, would only spread panic among the public. Forecasters weren't allowed to study tornadoes, or even to acknowledge their existence in public. That all changed after the Tri-State Tornado (see page 80).

Most funnel clouds do not make contact with the ground, but when they do, they officially become tornadoes. Witnesses describe the sound of a funnel cloud touching ground as "buzzing bees" or the roaring sound of a nearby waterfall. Residents in "Tornado Alley" know to seek immediate shelter the moment a funnel cloud is spotted, even if it has yet to touch ground.

There's never been another tornado like it. On March 18, 1925, the Tri-State Tornado rode a straight-line path for three and a half hours across 219 miles of Missouri, southern Illinois, and Indiana, making it the longest single tornado track anywhere in the world. With a mile-wide diameter, it looked wider than it was tall—so big that it lacked the classic genie-in-the-bottle funnel. By the time it had finished its run it had obliterated whole towns, sent thousands to the hospital, and killed more than twice the number of people lost in the next deadliest tornado in U.S. history.

The forecast for the central Midwest that day was for showers and cooling temperatures—nothing out of the ordinary. The weather bureau had been tracking a cold, low-pressure jet stream that bent down from western Canada into Wyoming and all the way to the Oklahoma-Texas border before curving back toward southeastern Missouri. Its wind speed was probably very strong, given how fast the tornado traveled.

A warm front from the Gulf of Mexico raised temperatures by 10 degrees in the region, causing warm air to rush skyward and providing what forecasters today call the tornado's lifting mechanism. The merged storm system transformed into a tornado-producing spiral, and the gray skies drizzling over southeastern Missouri began to turn an ominous black.

At 1 PM, a column of twisting air materialized near the town of Ellington, Missouri, killing a local farmer. Traveling 72 mph, it took fourteen minutes to hit Annapolis, where it killed four more people and destroyed 90 percent of the town. An hour later, the tornado spun off two adjacent funnels before it mowed through Biehle. It reformed into a single system a few miles later while making its way across the rolling farmland toward the Mississippi River. The tornado left Missouri; it had traveled 80 miles, mainly through rural farmland, and killed between eleven and thirteen people in 83 minutes. Illinois would fare far worse.

At 2:26 PM, the storm approached the town of Gorham, Illinois, raining down golf-ball-size hail as it advanced. "There was a great roar," recalled Judith Cox in the *St. Louis-Post Dispatch* two days later. "Like a train, but many, many times louder. 'My God!' I cried. 'It's a cyclone, and it's here.'" Cox was having lunch in town at the time. She tried to leave the restaurant, but was blown back inside as the building collapsed. The restaurant's cook was crushed, but Cox was pulled out of the rubble alive, along with a cow that had been picked up and dropped on the roof. The tornado killed thirty-seven of the town's five hundred residents and destroyed every building in it. Next in line was the twelve-hundred-person rail hub of Murphysboro.

The restaurant's cook was crushed, but Cox was pulled out of the rubble alive, along with a cow that had been picked up and dropped on the roof.

Wallace Akin, a retired tornado scientist from Drake University in Iowa, was two years old when the tornado closed in on his Murphysboro home. "An invading army of debris swept over the western hill—trees, boards, fences, roofs. Day became night," writes Akin in his 1992 history of the Tri-State Tornado, *The Forgotten Storm*. Akin's mother huddled over him in a family room as the tornado struck. "The house began to levitate and, at the same time, the piano shot across the room, gouging the floor and carpet

An overturned house was carried more than 50 feet from its foundation after a tornado touched down in Griffen, Indiana. The Tri-State Tornado was the deadliest tornado in U.S. history and was responsible for nearly 700 fatalities. Its 219-mile path of destruction remains the longest ever recorded. The tornado was part of a larger outbreak that affected Tennessee, Kentucky, and Indiana, as well as some southern states. By the time it was over, the tornado had obliterated entire towns, sent thousands to hospitals and killed more than twice the number of people lost in the next deadliest twister in U.S. history.

where I had played only moments before…. But the walls and the floor held as we and the house took flight."

Akin, his mother, and their entire home landed on top of the garage, which had landed on top of another house. Their once-tidy street was a wasteland of debris and houseless foundations. As the tornado closed in on downtown, Wallace's father was at his auto dealership.

"He went outside to chase down a car that was rolling away," Akin said. "Some type of debris hit him in the back of the head. His eyes popped out of their sockets, but the person who found him put them back in, and took him to the local hospital."

At the Longfellow Grade School, children were rushing out of the building as it collapsed, trapping roughly half of the 450 students. A block away, at the railroad repair yard, 35 men were killed as the tornado laid waste to their shops. According to Akin, the shops' many survivors rushed toward the school and began removing the rubble, "literally tearing their hands to the bone in their efforts." Eleven died there, and many more were hurt. Many of the seriously injured, including Akin's father, were put aboard an emergency train that left for St. Louis three hours later.

Rescuers worked for days to clear rubble and look for survivors in Murphysboro. At a local school, a young survivor was pulled alive from the rubble two days later. But fires also ripped through town in the wake of the tornado, and in one neighborhood a man trapped under rubble died when rescuers were forced to flee the approaching flames.

By Akin's estimate, 243 people died in Murphysboro that day, 463 people required surgery, and many had limbs amputated even as the supply of anesthetics ran out. Hundreds of others had minor injuries. The tornado depressed the town's economy for twenty years. The railroad repair center moved to Alabama, and then the Great Depression hit. It wasn't until World War II, Akin observed, that all the abandoned lots had been rebuilt and the town began to prosper again.

Joe Schaefer, the director of the National Weather Service's Storm Prediction Center,

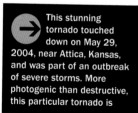

This stunning tornado touched down on May 29, 2004, near Attica, Kansas, and was part of an outbreak of severe storms. More photogenic than destructive, this particular tornado is unique for how the sun has peeked over the lower third of the laminar funnel.

THE FUJITA SCALE

In 1971, Ted Fujita and Allen Pearson developed a system to rate the intensity of tornadoes and called it the Fujita scale. In 2007, the Fujita scale was replaced by the Enhanced Fujita scale (EF scale), shown here.

EF NUMBER	Wind gust speed (in mph)
EF 0	65–85
EF 1	86–110
EF 2	111–135
EF 3	136–165
EF 4	166–200
EF 5	200 +

referred to the March 1925 phenomenon as the Murphysboro Tornado—and with good reason. The tornado claimed its greatest single disaster there. But the twister had another two hours and 130 miles to go before it petered out. In the mining town of West Frankfort, Illinois, miners were forced to climb out of a 500-foot underground shaft after the tornado cut power in the area. When they reached the surface, they found their community had been blown to splinters. Of the 127 people dead, most were women and children.

As the vortex crossed into Indiana at 4 PM, it hit its peak average speed of 73 mph and completely obliterated the town of Griffin, killing 25 and injuring 202. The tornado then veered off its straight-line course by 9 degrees. At that point, witnesses reported three separate funnels swirling within the larger maelstrom. The tornado was on a collision course with the town of Princeton, Indiana, and when it hit eighteen minutes later, it had lost none of its power. It destroyed 25 percent of the town and killed 45 people. The tornado rolled on for another twelve minutes before disintegrating into the history books. By then, it had killed 695 people, injured thousands more, and demolished 15,000 homes.

The Aftermath

"The single biggest thing that happened as a result of the Tri-State Tornado was the increase in public awareness about tornadoes," said Harold Brooks of the National Oceanic and Atmospheric Administration's (NOAA) National Severe Storms Laboratory in Norman, Oklahoma. Even though there had been a ban on using the word *tornado*, Brooks noted, it was the beginning of local tornado-spotter networks. "There were no official programs that we know of, but when you look at old newspapers you start to see mention of these spotters after 1925." According to Brooks, the storm-spotter programs contributed to a steady decline in the number of tornado deaths in subsequent years. Today, fifty people are killed by tornadoes annually. If things had continued at the 1925 rate, that number would be closer to five hundred.

It took another twenty-three years before modern tornado forecasting was born. On March 20, 1948, a tornado tore through Tinker Air Force Base near Oklahoma City. General Fred Borum asked two of his meteorologists, Captain Robert Miller and Major Ernest Fawbush, to look into predicting tornadoes for Tinker. The forecasters' first major test came just five days later, when they informed Borum that the weather looked a lot like it had on March 20. Borum asked if they were going to issue a tornado forecast.

With much trepidation, Fawbush and Miller issued their alert for between 5 and 6 o'clock that night. They were right: a tornado once again tore through Tinker. Not long after that, the U.S. Weather Bureau dropped its ban on the word *tornado*.

ANATOMY OF A TWISTER

A tornado forms when three drastically different masses of air collide (hot, dry air; warm, moist air; and cold, dry air). Because of the unique topography of the U.S. (the Sonoran Desert to the west; Gulf of Mexico in the south; and the Rockies and Canada to the north and west), it's prone to spouting off twisters. These masses of air crash together over the flat Midwest region, where the atmospheric instability forms super cells of thunderstorms.

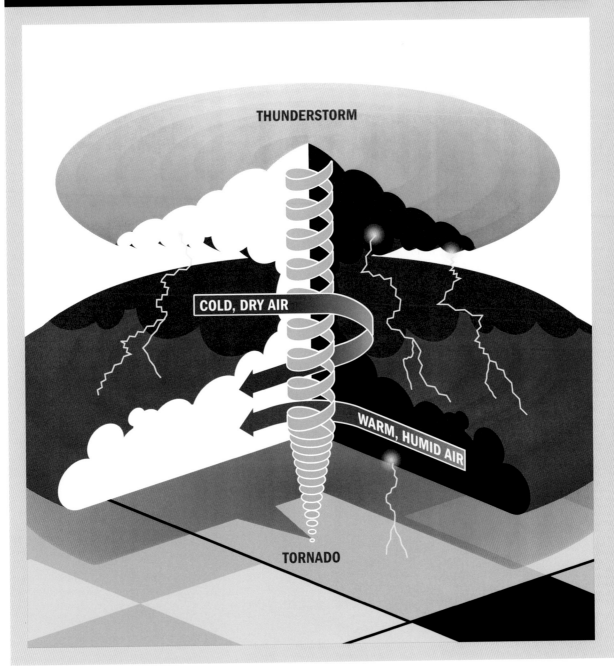

THUNDERSTORM

COLD, DRY AIR

WARM, HUMID AIR

TORNADO

A house on U.S. 42 north of downtown Xenia, Ohio, was cut in half after a tornado touched down on April 3, 1974 and devastated the town. The tornado measured F5 on the Fujita scale and was part of the Super Outbreak—the largest series of tornadoes recorded in history. Half of Xenia's buildings were destroyed in the tornado, which killed 34 people.

THE SUPER OUTBREAK OF 1974

A Textbook Formula for Tornado Devastation Across 13 States

On that fateful April 2, a sprawling mass of cold, dry air dropped down from Canada toward the Mississippi and Ohio River valleys, and an opposite mass of warm, moist air pushed northward from the Gulf of Mexico. They were set to converge beneath an intense jet stream with 140-mph winds at an altitude of 40,000 feet. Even using primitive 1950s-era radar, National Weather Service forecasters could tell something bad was brewing over the central United States.

As the three patterns collided, the contrast between the cold and warm fronts caused the warming air to rush upwards. The winds spiraled counterclockwise, and the jet stream propelled everything forward. It was a textbook formula for tornado creation.

On April 3, all hell broke loose: the storm system spun 148 tornadoes across thirteen states. The first twisters touched down at 2 PM in Bradley County in south-central Tennessee and Gilmer County in northwestern Georgia. Ten minutes later and 450 miles away, tornadoes were loosed across Central Illinois and across Indiana ten minutes after that. Twisters raged as far south as Laurel, Mississippi, as far north as Detroit, and all the way east to Staunton, Virginia. A state trooper in Martinsburg, Indiana, reported that the town just disappeared. After fifteen hours, the tornadoes finally died down. They had traveled 2,014 miles; the longest twister wreaked havoc for more than 100 miles.

The power of the twisters during the Super Outbreak of 1974 was truly epic. Until 2007 (see sidebar on page 84) tornado intensity was measured by the Fujita scale, which ranged from an F0, with winds as low as 40 mph, to an F5, with winds that reaching more than 300 mph. In 1973, an active tornado year, eleven hundred twisters were reported across the United States—but only one measured F5. During the Super Outbreak, however, six tornadoes measured F5 on the Fujita scale, and another eighty-eight reached the 113- to 157-mph wind speeds of an F2. "There's never been anything like it before, as far as we know, or since," says Joseph Shaefer, director of NOAA's Storm Prediction Center. "It was a once-in-a-century event, and probably rarer than that."

The most powerful twister of the outbreak was also the deadliest. At 4:30 PM, three separate storms converged to form a single tornado that landed on a house owned by the Winston family, 10 miles outside Xenia, Ohio, and 50 miles southwest of Columbus. Debby Winston, seventeen years old at the time, saw the giant black cloud approaching and dove into a closet with her mother and younger sister just before their house exploded. Winston, who told her story to the *Akron Beacon Journal* on the twenty-fifth anniversary of the storm, awoke 100 yards from her home with a gash under her eye. She fared better than her sister, who punctured her hip, and her mother, who broke her collar and pelvic bones.

Traveling northeast at 50 mph, the tornado then made a beeline for Xenia. It had one of the strongest wind speeds ever recorded: 318 mph. (Meteorologists later debated whether to rate the tornado an off-the-charts F6.) It was so large that people on the ground couldn't even see a funnel cloud—just a black, swirling, half-mile-wide maelstrom. Jeff Louderback was just five years old at the time, but the memory of riding out the storm in his house remained vivid decades later. He described it to the Associated Press in 1999: "Mom and Dad covered me, shielding my body from flying bricks and shattered glass. The deafening wind sounded like a team of fighter jets. I saw bedroom doors slamming against the wall before flying off their hinges. The roof ripped off, and the walls around us crumbled. Between my sobs, I could hear Dad praying for our protection."

People couldn't even see a funnel cloud—just a black, swirling, half-mile-wide maelstrom.

Betty Hill was in her home in the Arrowood subdivision of Xenia when the tornado approached, and she piled into a bathtub with a neighbor and four children. "The sound was like a huge semi coming up a long valley," Hill told Cleveland's *Plain Dealer* in March 1999. "We stayed in there until we couldn't hear it anymore, then went out front to take a look." Hill saw the tornado rip the roof off the local high school as it moved toward downtown.

Firefighter Charles Beason watched the tornado's approach with his binoculars. "That cloud just kept getting bigger and bigger and bigger. It was so huge and so close to the ground that I thought, Oh my God!" he told *The Plain Dealer*. "Every once in a while I could see a roof go up, start spiraling, then just explode into thousands of pieces."

Semi trucks were picked up and dropped atop the local bowling alley. Train cars went flying, as did school buses, which eventually landed atop the destroyed school. There were many tragic fatalities: a young child sucked into the tornado, five people crushed under the debris of the local A&W restaurant. Yet there were also seemingly miraculous stories of survival: the tornado shattered hundreds of glass plates in the Xenia glass shop with five children inside, but the razor-sharp shards were vacuumed right out into the vortex.

The tornado wreaked havoc in Xenia for nine long minutes before finally heading out of town. On the way, it picked up a herd of cows owned by Mildred and Howard DeHaven and spun them into another field. Two of their horses ended up in neighbors' trees.

By 4:50 PM the tornado was gone, but half the town's buildings were destroyed, including seven schools, nine churches, and more than three hundred homes. The local hospital quickly became inundated by survivors with broken limbs. The tornado caused an estimated $100 million in damage in Greene County, where Xenia is located. Nationally, the total reached $600 million. And while Xenia's death toll was the single largest, it accounted for just 10 percent of the outbreak's 330 fatalities. For another thirteen hours, tornadoes continued to destroy homes, farmland, and lives across the central and eastern United States.

The Aftermath

"Two very important events came from the 1974 outbreak," says Harold Brooks, a research meteorologist with NOAA's National Severe Storms Laboratory. "First, the National Weather Service adopted the Fujita scale. And second, support and money for tornado-intercept operations greatly increased."

Ted Fujita, a meteorologist at the University of Chicago, developed his eponymous scale to rate the power of tornadoes in 1971, but his system took hold in 1974, when he catalogued the entire 148-tornado outbreak. "It might seem minor," Brooks says, "but it wasn't. It allowed the entire scientific community to use the same language and the same framework for studying tornadoes." Instead of relying on the misleading size of a tornado (small tornadoes can pack ferocious winds, while large ones can be weak), the Fujita scale is based on wind speed, which has proven to be a much more accurate predictor of a tornado's potential to cause damage.

Likewise, sending scientists out into the field to "chase" tornadoes yielded its own advances. "Before intercept operations, we didn't even know what part of a storm a tornado came from," Brooks said. They typically form on the back right side of a storm, something that couldn't be seen on radar.

"Now we have a better idea of where a tornado will strike and whom we need to warn."

Since 1974, the National Weather Service has gone from 52 weather stations to more than 120. Each station employs at least one meteorologist to keep an eye on approaching storms and another to inform the public and emergency officials about developing problems. Together with the National Oceanographic and Atmospheric Administration, it has also spent more than $4.5 billion upgrading forecasting technology.

Immediately after the Super Outbreak of 1974, Congress funded a techniques development unit at the National Weather Service. In 1979 the group superimposed real-time satellite images over digital maps for the first time. This technique developed into the Advanced Weather Interactive Processing System, now used across the country, which integrates satellite, radar, weather observations, and numerical model forecasts.

The biggest single advance has been vastly improved radar. In 1974, the National Weather Service still used 1950s-era vacuum-tube radars that bounced electromagnetic waves off approaching storms. They could pinpoint the position of a storm, but couldn't tell which direction it was headed or what was going on inside it.

Doppler radar, which the military used during the Cold War to scan for incoming Soviet missiles, measures the change in frequencies after a radar wave bounces off an approaching object, and uses the variation to calculate speed and direction, measure rainfall, and search for signs of spiraling tornadic winds. By 1977, NOAA's Severe Storms Laboratory had a research radar up and running, but it wasn't until 1991 that Congress approved a nationwide Doppler radar system. Six years and $600 million later, 170 Doppler radars could scan the entire country for bad weather. Tornadoes can now accurately be predicted with an 80 percent success rate.

"It has dramatically changed the warning times for tornadoes," says Don Burgess, a thirty-two-year veteran of the Severe Storms Lab. "In the seventies, the lead time was about zero. Now the average lead time is twelve to fourteen minutes. It doesn't seem like a lot, but when you need to take shelter, every minute counts."

The scariest thing about tornadoes, as well as what makes them so difficult to learn about, is the speed with which they appear and disappear. Those arriving on the scene within mere minutes of a tornado are usually far too late to study them. With diligence and commitment to the development of early warning technology and scientific advances, we will continue to study and survive these awesome and deadly phenomena.

A tornado-stricken family in Xenia, Ohio, rummages through what was their home, attempting to collect items of sentiment value—in this case, a scrapbook and an oil lamp. 10,000 people in Xenia were made homeless in the wake of the tornado, and President Richard Nixon came to visit after the town's plight made headlines around the world.

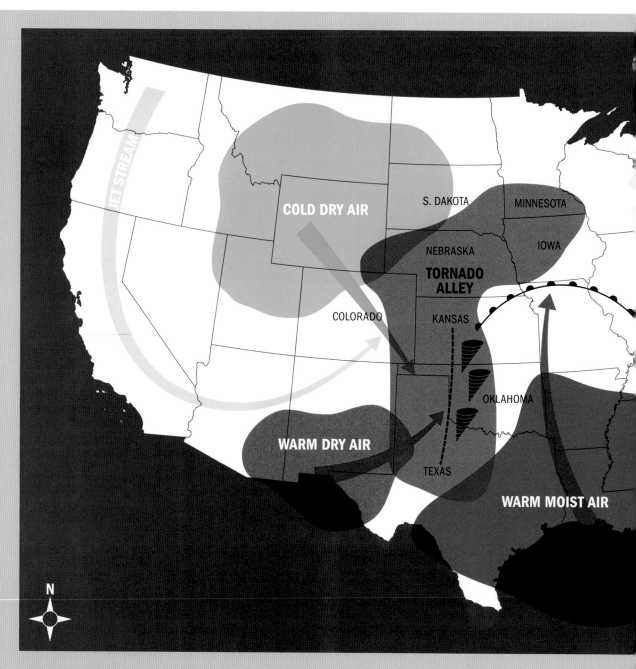

According to the National Climatic Data Center, the states with the most tornadoes are as follows (cumulative between January 1, 1950 and July 31, 2009):

State	Tornadoes	State	Tornadoes
Texas	8,049	Iowa	2,368
Kansas	3,809	Illinois	2,207
Oklahoma	3,442	Missouri	2,119
Florida	3,032	Mississippi	1,972
Nebraska	2,595	Alabama	1,844

TORNADO ALLEY, USA

The majority of tornadoes in this country form in a part of the central United States that has been dubbed Tornado Alley by the media. It's an area roughly defined by the plains between the Appalachian Mountains to the east and the Rocky Mountains to the west. (In essence, it spans from western Ohio in the east to central Kansas and Nebraska to the west, and stretches from central Texas in the south to Iowa in the north.) Though the National Weather Service has not officially defined the colloquialism, there is no questioning the fact that 90 percent of U.S. tornadoes appear in this region of the country.

The reason tornadoes strike Tornado Alley so often is because it is where the hot, dry air from the Sonoran Desert collides with warm, moist air from the Gulf of Mexico and cold, dry air from Canada and the Rocky Mountains. When these air masses crash into one another over the flat land of Tornado Alley, the atmospheric instability forms a front that can lead to strong storm systems (or super cells) of thunderstorms and heavy precipitation—and oftentimes, a tornado outbreak.

Over the last half century, Texas (due largely to the size of the state) has experienced the most tornadoes, averaging more than one hundred each year. Kansas and Oklahoma are not far behind. In fact, the building codes in Tornado Alley are stricter than those in many other states, requiring more secure roofing and connections to the foundation of buildings. Houses often have storm cellars, tornado sirens, and weather alert systems because of the severity of the tornadoes that haunt this region.

SURVIVAL TIPS: TORNADOES

Thousand find themselves in the following deadly situation each year, especially in Tornado Alley. They are driving home from work when they spot a dangerous, funnel-shaped cloud. Instinctively, they speed away in the other direction, but the tornado keeps gaining. Here's what everyone should know:

- If you can't outrun the twister, get out of the car. Vehicles are nothing more than half-ton missiles in the path of a tornado. Avoid seeking shelter beneath an overpass; they're not safe from a twister and create dangerous traffic conditions during a panic.

- Get inside. The basement of a house generally provides the best shelter during a tornado. If you're in a dorm, apartment, or office, get to the lowest floor. There is no basis to the myth that the southwest corner of a basement is the safest; it's just as dangerous as the other three.

- Avoid the windows. Tornadoes blow spears of glass from windows inward. Another myth is that opening windows will lower the pressure in a building during a tornado: doing so is a waste of time. Find a central room and cover yourself with a mattress instead.

A tornado's funnel is a thin tube that reaches from the cloud to the ground. A translucent dust cloud, which is kicked up by the tornado's strong winds near the ground, has a much wider radius than the rest of the funnel. Wind speeds can reach more than 300 mph and travel dozens of miles.

Hurricane Andrew wasn't the only major hurricane of the 1992 Atlantic hurricane season, but it caused more than $40 million (2010 USD) in damage, making it the costliest hurricane in U.S. history until Hurricane Katrina. This time-lapsed photograph shows Hurricane Andrew moving east to west on August 24, 1992, and August 25, 1992.

HURRICANES

Hurricanes form over bodies of warm water when, put very simply, warm air rises and collides with cool air. Such a system is first called a tropical disturbance; it may become a tropical depression if its clouds and thunderstorms become defined and organized. A tropical storm is born when winds in such a system reach more than 39 mph and a distinct cyclonic shape begins to form. A tropical storm becomes a hurricane when its wind speeds reach 73 mph and an eye becomes visible.

The year 2005 was especially horrific for hurricanes. Though hurricanes Rita and Wilma would surpass it in terms of power, it is Hurricane Katrina, and the ensuing flooding of New Orleans, that remains etched in everyone's mind. Its warm, wet climate and its proximity to the Atlantic Ocean and the Gulf of Mexico make the American South particularly susceptible to hurricanes, but it is not always the South that feels the fury of one of nature's deadliest events.

THE NEW ENGLAND HURRICANE

Long Island, New York, on September 21, 1938

As northeasterners buttoned their jackets against the chill of early fall, a hurricane brought the driving rain, battering salt spray, and lethal winds of the tropics to New York's Long Island. The powerful storm overcame immense obstacles—cold water, east-moving ocean currents, sea-bound winds—to reach the dense, vulnerable northern cities and seashores, where it left catastrophe in its wake.

It had been more than a century since such a powerful hurricane had hit New England (the previous one was in 1815). During normal summertime weather patterns, the east-flowing winds of the jet stream push impending northeast hurricanes out to sea. But on that September day, according to Harvey Thurm, hurricane program manager for the eastern region of the National Oceanic and Atmospheric Administration (NOAA), reconstructed air maps suggest that the jet stream's flow bent southward toward the Gulf of Mexico and then blew northward up the eastern seaboard. "Once the hurricane got caught up in that, it was like a cork in a river's flow," Thurm says.

But even that would not have been enough to propel the storm up the coast. Warm water acts as fuel for hurricanes, and above Cape Hatteras in North Carolina, where the warm ocean currents of the Gulf Stream veer east, the water temperature begins to drop. "It's got to be a very powerful, very fast-moving hurricane to make it through the cold water," Thurm adds. And it was: traveling at 60 to 70 mph (twice as fast as many hurricanes) and

The hurricane's relentless waves punched a hole through the beach at Shinnecock Bay, cutting it in two and creating Shinnecock Inlet, which still exists today.

aided by high-pressure systems to the east and west that formed a canal of low pressure, the hurricane muscled its way past Cape Hatteras, plowed through the cold water, and barreled into Long Island in just six hours.

Residents were caught off guard when, at 3:30 PM on September 21, a hulking, foaming 16-foot storm surge crashed ashore, spreading two-story waves from central Long Island to Gloucester, Massachusetts, about 250 miles north. The surge enveloped shore towns, leaving standing water 6 feet high and killing twenty-nine in the small community of Westhampton, New York, alone. Survivors "grabbed onto telephone poles as boards, planks, and timbers pelted them," writes Everett Allen in *A Wind to Shake the World*, a book about the storm. "The wind was southeast at well over 100 mph, and the sand in the air was terrific." The troughs of towering waves left the bay floor dry, and the wind drove ducks backward in flight.

While the storm focused its force on the shore, waves littered the interior of Long Island with debris. The water tossed dozens of boats 300 feet in from the coast, destroyed hundreds of houses, washed wreckage out to sea or deposited it inland, and scattered cars everywhere. The hurricane's relentless waves punched a hole through the beach at Shinnecock Bay, cutting it in two and creating Shinnecock Inlet, which still exists today.

After devastating New York's coast, the hurricane's eye skipped across Long Island Sound and headed straight for Connecticut, Rhode Island, and Massachusetts, hitting each with equal fury. Winds blew as high as 120 mph, mercilessly driving the storm surge into the shore.

In New London, Connecticut, at the mouth of the Thames River, the storm surge took the *Marsala*, a colossal, 300-foot, five-masted cargo ship, along with its 8- and 10-ton anchors, and

Streets of the quiet New England town of Winchedon, Massachusetts, became raging torrents as heavy rains overflowed nearby rivers after the New England Hurricane of 1938. The flood reached its climax on September 21 when the hurricane blew down countless structures and added torrential rains to the inundated areas. More than 400 people were killed and thousands injured.

slammed it into the town's dockside warehouses, sparking a fire.

To the east, in Rhode Island, residents peered toward the ocean at what appeared to be a fast-moving fog bank. But when it drove closer, they realized that they were coming face-to-face with 40-foot waves pushed by the storm surge. One hundred people were killed as the hurricane came ashore around Westerly and swept five hundred homes into the sea.

The surge then worked its way north and west, funneling into Narragansett Bay and overwhelming the state's capital, Providence, at 4:45 PM, just a little over an hour past its first landfall. Some Providence citizens retreated to the steps of the city hall as 14 feet of water toppled over vehicles, trolley cars, and pedestrians. At the Biltmore, hotel guests hung knotted sheets out of their windows so that others caught in the flood below could grab them. At least 262 people were killed in Rhode Island, though in the 2003 book, *Sudden Sea: The Great Hurricane of 1938*, author R. A. Scotti contends that the state's death toll was closer to 433. According to Scotti, plane crews performed the final tally by air, counting heads bobbing in the water.

The storm surge came ashore as far east as Massachusetts's Cape Cod, where 8 feet of water covered the resort town of Falmouth. Meanwhile, the eye of the hurricane headed straight up the Connecticut River and dropped up to 17 inches of

Raging waters made transportation impossible as sections of highway along the New England Coast were flooded. The Connecticut River rose an astonishing 20 feet, ascending the banks near the capital city of Hartford, CT. Ninety-seven people were killed in the state.

rain as it pushed northward toward Canada. The ground in western New England was already saturated from several days of rain, contributing to flood conditions. The Connecticut River rose 20 feet, burst its banks near the state's capital, Hartford, and spread out 4 miles. Ninety-seven people died in Connecticut.

By 8 PM, not quite twelve hours after the hurricane passed Cape Hatteras, the storm mowed through interior Vermont, toppling tens of thousands of trees and uprooting whole apple orchards in Burlington. In Montpelier, miles from Long Island Sound, the hurricane lashed houses with the salt spray it had picked up in the Atlantic Ocean.

The hurricane crossed into Canada by 9 PM and faded away, leaving $6.2 million dollars in wreckage (or more than $15 billion in 2006 dollars) in its wake. In addition to killing nearly 700 people, the hurricane and the resultant flooding and storm surge injured 3,500, left 63,000 homeless, damaged 75,000 buildings, and sunk 3,000 boats as it came to shore.

The Aftermath

After more than a hundred years without a hurricane, the 1938 storm reminded northeasterners that they, too, are vulnerable to tropical-force destruction. And because a storm must be incredibly fast and strong to reach New England, when a hurricane does come, it comes hard.

Despite obstacles such as the jet stream and the Gulf Stream, five hurricanes have hit the northeast coast since 1938. The last one was Gloria, in 1985. Though none have connected with New York City, a direct, powerful hit could easily become the most catastrophic hurricane in U.S. history because of the city's high population density and the difficulty of evacuating the island of Manhattan through its limited network of bridges and tunnels.

It took decades for NOAA and the National Weather Service to create the technology that allows them to predict the path of hurricanes and warn the vulnerable and unsuspecting—exactly what didn't occur in 1938. Primitive radars came on line in the 1950s, but the more sophisticated Doppler radar wasn't operational across the United States until the 1990s. Today, forecasters also use satellites and computer modeling to predict a hurricane's path and warn the public before it's too late.

To help the National Weather Service make its hurricane predictions, NOAA also operates a small air force of planes that fly into, around, and in front of hurricanes. "Besides the hurricane hunters that fly into the eye of the hurricane, there's an NOAA Gulfstream G-IV that flies at a high altitude, sampling the storm environment ahead of a hurricane," Thurm said. "It's a whole new world compared to 1938."

ANATOMY OF A HURRICANE

Hurricanes build their energy as they glide across the ocean, pulling up moist, warm tropical air from the surface and ejecting cool air above it. The calm, low-pressure center or eye is surrounded by the eye wall, which contains the hurricane's most violent winds.

Hurricanes spin counter-clockwise in the Northern Hemisphere and clockwise in the Southern Hemisphere (due to the Coriolis effect). Warm, humid air rises and spirals toward the eye to form the clouds of the storm, and light winds outside the hurricane guide its direction and enable the hurricane to gain strength.

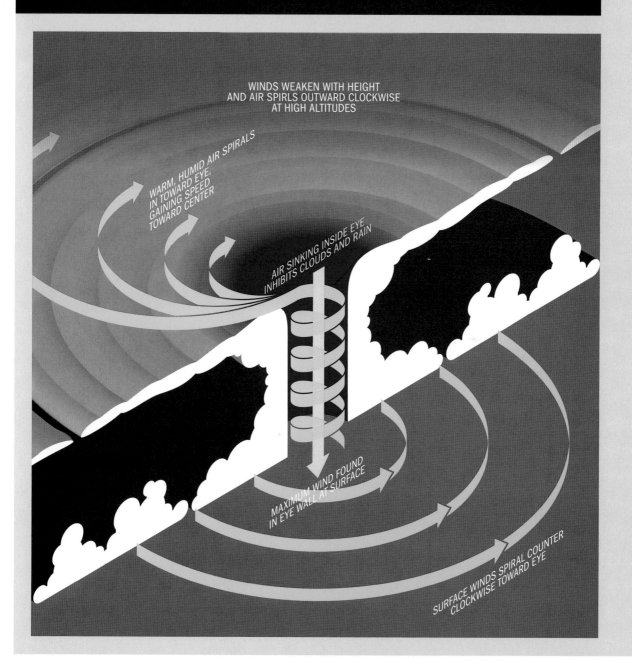

WINDS WEAKEN WITH HEIGHT
AND AIR SPIRLS OUTWARD CLOCKWISE
AT HIGH ALTITUDES

WARM, HUMID AIR SPIRALS IN TOWARD EYE, GAINING SPEED TOWARD CENTER

AIR SINKING INSIDE EYE INHIBITS CLOUDS AND RAIN

MAXIMUM WIND FOUND IN EYE WALL AT SURFACE

SURFACE WINDS SPIRAL COUNTER CLOCKWISE TOWARD EYE

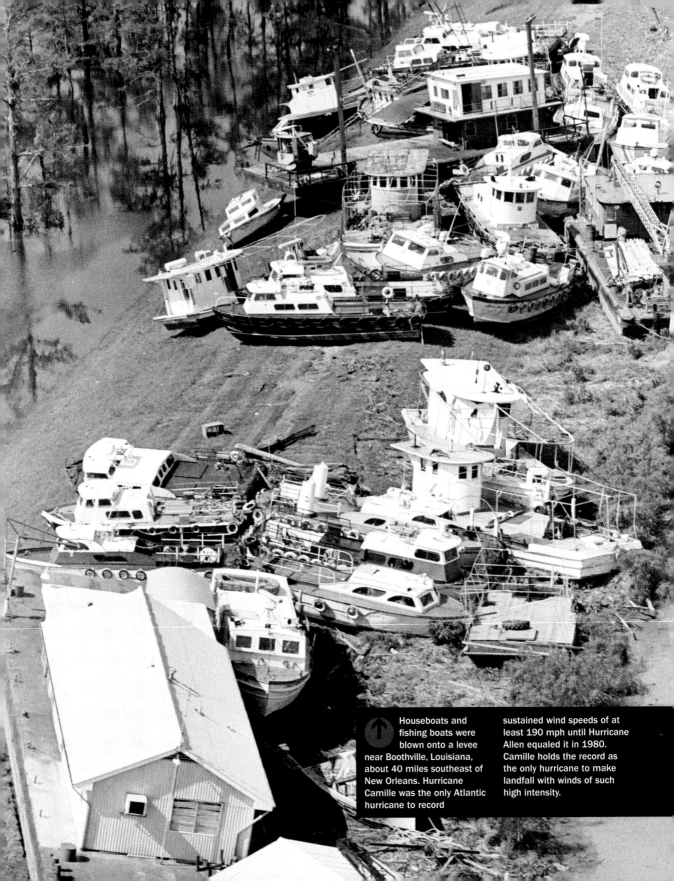

Houseboats and fishing boats were blown onto a levee near Boothville, Louisiana, about 40 miles southeast of New Orleans. Hurricane Camille was the only Atlantic hurricane to record sustained wind speeds of at least 190 mph until Hurricane Allen equaled it in 1980. Camille holds the record as the only hurricane to make landfall with winds of such high intensity.

Hurricane Camille was the second most intense storm to reach the United States in the twentieth century, trumped only by the Labor Day hurricane at the Florida Keys in 1935. The wind roared down from the Mississippi hills and emptied rivers and swept bayous dry. Three hours later, at 8:00 PM, the wind reversed itself. The Gulf of Mexico rushed back. By 9:00 PM four cities along 50 miles of Mississippi's lovely "chain of pearls" coast—Pascagoula, Biloxi, Gulfport, and Pass Christian—were being blown apart.

This was a wind. National Weather Bureau officials called it the worst storm in history. They named it Hurricane Camille.

But Camille was not a typical hurricane: it was a compact bomb. Hurricanes are usually hundreds of miles across, but Camille was only about 50 miles wide. It was like a supertornado that spawned many lesser tornadoes.

Navy, air force and weather bureau planes sped out to meet the monster that had come up from Cuba overnight. "We've got 50 knots of wind out here," radioed an air force crew. Soon shore-based radars were picking up a twister of a cloud 10 miles tall. It glittered in the morning sun, 200 miles west of Florida. Then it steamed north, a juggernaut aimed at Mississippi's sandy coast.

On Sunday at 4:30 AM, Camille was still pushing north. That's when Bill Tilson, head of the Mobile, Alabama, weather bureau office, sent messages alerting the Mississippi coast: the storm is coming your way. Get out!

At 8:00 AM Paul Hearon, owner of an Amoco gas station in Biloxi two blocks from the sea, decided he'd better move his tires to higher ground. It was slow work because people kept buying white gasoline for lanterns and stoves. Everyone knew that the electrical power would fail. At noon he locked up and headed for the hills.

On U.S. 90, behind the yacht club, Mr. and Mrs. Bill Hood checked out fifty-three scared guests from their T-shaped Towne House Motel. Seven other guests lingered, but decided to move upstairs into Rooms 11 and 12, close to U.S. 90. At noon, a young couple came to the motel. "We have no car to escape in and no place to stay," they told Hood. "May we come in?" Hood gave them Room 16 on the second deck.

One-half mile east, at the armory, one hundred young National Guard trainees nervously checked the rescue boats they might need later.

Twenty-five miles west of Biloxi, at Pass Christian, Civil Defense Director Parnell McKay began urging people who had not already escaped to move into two big schools. McKay asked Police Chief Gerald Peralta to try to drive to the swank, three-story Richelieu apartments and urge the twenty-four occupants still there to go inland because the Richelieu lay only 750 feet from the Gulf.

"We went to each door and knocked and asked everybody to get out while they still could," said Peralta. About half complied. Then he radioed to the firehouse and told McKay: "The others insist on staying."

At 5:00 PM the eye of the storm, surrounded by winds of over 200 mph, was still 90 miles south of the coast and passing the mouth of the Mississippi, where the *LST*, an offshore drilling tender, was anchored. In addition to its anchor, the *LST* was secured by a line to a huge tree on shore. Then the wind and waves hit the ship with such force that it dragged its anchor and pulled the tree over. The ship was propelled backward, out into the Gulf.

The scream of the wind rose. Then came the collision. The *LST* was hit by two unseen ocean freighters, and all three ships drifted off together.

The alarm to man the rescue boats reached the Biloxi National Guardsmen when the wind

"...go back and tell them I want the names of next of kin for everybody there."

suddenly reversed itself toward 8:00 PM and the tide came rushing in. The men sloshed through knee-deep water to board six landing craft called LARCs—light amphibious reconnaissance crafts. They rumbled toward town. The wind was 80 mph and growing stronger.

It was about 9:00 PM when screams rose above the noise of the wind at the Towne House Motel. Owner Bill Hood knew the couple in Room 16 was in trouble. But to get outside would be tough. U.S. 90 was deep in ocean.

And then Hood heard the tornadoes. "I've heard them before and I know the sound," he said. The west sky turned red with electrical fire. The big Pontiac garage next door exploded. Hood saw the

Biloxi Yacht Club spin around. Then it blew away like paper before an electric fan. A yacht whirled ashore.

Screams still came from Room 16, and there were also thuds—furniture floating against the ceilings downstairs. It was almost impossible to open the door against the wind. When Hood and the air force lieutenant who volunteered to accompany him got out on the porch, it was like being on a ship's deck in a storm.

Waves crashed over the railing and smashed them against the wall. They crept from door to door. It took twenty minutes to reach No. 16—the fourth room down. When the door wouldn't yield, they smashed a window and climbed in.

The girl was hysterical. Her husband, hands bloodied, was trying to smash a hole through the wall. Hood and the lieutenant grabbed them and led them, one at a time, back to the other rooms.

Incredibly, the young guardsmen in the LARCs kept going through it all. They bumped submerged cars, tangled with wires, got blown toward the Gulf, and struggled back to tie up to rooftops. They saved hundreds of people.

Back in Pass Christian, a group of men and boys were rescuing people from treetops in an amphibious military personnel carrier (DUCK) borrowed from the National Guard. At 10:00 PM the DUCK got a frightening radio message: the Episcopal church had blown away with fourteen people in it. Soon afterward, Civil Defense Director Parnell McKay also got a call: the Randolph School gym had just collapsed. Luckily, no one had been in the gym at the time. But McKay now was concerned about the seven hundred people in the public high school gym: if one gym had given way, another might, too.

He wasn't worried about the strongly built Catholic school gym. But he flashed a radio order

10 MOST DEVASTATING HURRICANES
(adjusted to 2008 USD)

Name	Date	Cost in Billions
1. Katrina	2005	$46
2. Andrew	1992	$24
3. Ike	2008	$13
4. Wilma	2005	$11
5. Charley	2004	$9
6. Ivan	2005	$8
7. Hugo	1989	$7
8. Rita	2005	$6
9. Frances	2004	$5
10. Jeanne	2004	$1

Service station owner H.A. Torgerson moves debris from what had been the town post office next door to his station in Waveland, Mississippi, on September 11, 1969. Hurricane Camille passed Cuba as a Category 3 storm, then strengthened over the Gulf of Mexico with a storm surge of 24 feet. The hurricane flattened nearly everything along the Mississippi coast and caused $1.42 billion ($8.43 billion in 2010 USD) in damages.

to the high school: "Get those people out of the gym—quick!" They came out, wading, stumbling, gasping in the wind. The last stragglers had just reached the main schoolhouse when, with a roar, the gym walls blew and its massive roof caved in. As many as five hundred lives may have been saved by McKay's quick thinking.

The Aftermath

By 2:00 AM, Monday, the worst was over. A total of 136 people had been killed along the Mississippi coast—over 100 in Gulfport, Long Beach, and Pass Christian, and at least 15 in Biloxi. (Three days later, Camille, swinging northeast as it petered out, unleashed furious rains over Virginia, killing another 100 people.) Property damage in Mississippi totaled nearly $1 billion. Over 36,000 homes and nearly 600 mobile homes had been destroyed or damaged. The Towne House Motel—or at least a good part of it—remained standing. But not a trace remained of the Richelieu apartments. The oil company ship *LST* miraculously survived.

The National Weather Bureau, though it did a good job of warning residents, was so shaken by all the death and destruction that it has wiped the name Camille off its lists forever—"retired" it like Babe Ruth's famous No. 3. Other names may be reused on future hurricanes. But never again will a storm be called Camille.

HURRICANE KATRINA

The Lessons Taught By New Orleans: August 29, 2005

No discussion of hurricanes would be complete without 2005's Hurricane Katrina, the largest natural disaster in the history of the United States. Five times as many people died as a result of Katrina than Camille. And we're still picking up the pieces.

"DEVASTATING DAMAGE EXPECTED."

By the time the National Weather Service issued this ominous alert on the morning of Sunday, August 28, 2005, Hurricane Katrina had morphed from a relatively weak Category 1 hurricane to a Category 5 tropical monster—and was spiraling straight toward New Orleans.

The city would be "uninhabitable for weeks… perhaps longer," the weather service warned. Half the houses would lose their roofs. Commercial buildings would be unusable, and apartment buildings would be destroyed. Residents should expect long-term power outages and water shortages that would "make human suffering incredible by modern standards."

In fact, the only outcome the 258-word alert didn't specifically foretell was the massive flooding that would leave most of New Orleans submerged under a fetid stew of water and chemical runoff. But the likelihood of that happening had been well-known for years. The city sits as much as 10 feet below sea level, between the Mississippi River and Lake Pontchartrain, and is kept dry by a complicated system of canals, levees, and pumping stations. Publications such as *Scientific American*, *National Geographic*, *Popular Mechanics*, and New Orleans's *Times-Picayune* had all reported on the city's vulnerability in the event of a major hurricane.

National Hurricane Center director Max Mayfield was so concerned about the potential consequences of Katrina that he called the mayor of New Orleans, the governors of Louisiana and Mississippi, and even President George W. Bush at his ranch in Texas. Mayor C. Ray Nagin of New Orleans issued a nonmandatory evacuation order by Sunday and, despite later misperceptions, most people left.

> "I know they're saying 'get out of town,' but I don't have any way to get out. If you don't have no money, you can't go."

Highway lanes were converted to outbound traffic only, and state police estimated that eighteen thousand vehicles per hour were streaming away from the region by late afternoon. Before Katrina hit, 80 percent of residents had already evacuated; but an estimated one hundred thousand people remained. Some officials argue that they stayed by choice, but reports on the ground suggest that many residents, and quite a few tourists, were simply stranded. Estimates vary, but between twenty-six thousand and one hundred thousand New Orleans families did not own a car. Greyhound and Amtrak cancelled service into and out of New Orleans on Sunday, and airlines grounded planes at New Orleans's Louis Armstrong Airport.

"I know they're saying 'get out of town,' but I don't have any way to get out," New Orleans resident Hattie Johns, 74, told the Gannett News

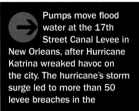

Pumps move flood water at the 17th Street Canal Levee in New Orleans, after Hurricane Katrina wreaked havoc on the city. The hurricane's storm surge led to more than 50 levee breaches in the federally built levee system that protects metro New Orleans. The 17th Street Canal, also known as the Upperline Canal, is a drainage canal that flows into Lake Pontchartrain.

WHEN THE LEVEES BREAK:

Why New Orleans Flooded After Katrina

Though many accounts portray Katrina as a storm of unprecedented magnitude, it was in fact a large, but otherwise typical, hurricane. On the 1-to-5 Saffir-Simpson scale, Katrina was a midlevel Category 3 hurricane upon its landfall. By contrast, meteorologists estimate that 1969's Category 5 Hurricane Camille, which followed a path similar to Katrina's, packed winds as high as 200 mph while Katrina's winds in New Orleans barely reached hurricane strength.

What made Katrina so devastating was that she approached over the Gulf of Mexico's shallow northern shelf, generating a more powerful storm surge—the water pushed ashore by hurricanes—than systems that move across deeper waters. In Plaquemines Parish, south of New Orleans, the surge topped out at 30 feet; in New Orleans the surge was 15 feet—enough to overtop some of the city's floodwalls.

There were more than 50 failures of the floodwalls and levees in and around New Orleans, as the city was flooded by millions of gallons of water that left parts of New Orleans in 10 feet of water. The illustration of New Orleans shows the city at the peak water levels directly after Katrina.

Most of the New Orleans flood barriers are simple earthen embankments, or levees, supporting a wall of steel sheet piles, some of which are capped with reinforced concrete I-walls. Breaches occurred when storm surges poured over the walls (as shown in illustration 1), washing away, or souring, interior soil foundations. This weakened their lateral stability and pressure from the floodwaters then caused collapse.

The cause of breaches on the Seventeenth Street and London Avenue canals remains a mystery. Over-dredging in the Seventeenth Street Canal may have removed lining sediments near the floodwall's sheet-pile wall, allowing water to percolate through deep levee soils. Swimming pools and other structures built too close to the barrier may have compromised its integrity by compressing its foundation, as shown in illustration 2. Later reports by the U.S. Army Corps of Engineers found that the floodwalls were so poorly designed that the maximum safe load was not even capable of holding half the original design of 14 feet of water.

Months after independent investigations concluded that the levee failures in Louisiana could not be attributed solely to natural forces that exceeded the original design's intended strength, the Corps of Engineers admitted that there were "problems with the design of the structure." The final report by the Interagency Performance Evaluation Task Force (IPET) that was assembled after Katrina states that the destructive forces of the hurricane were "aided by incomplete protection, lower than authorized structures and levee sections with erodible materials."

One idea to prevent future storm surges was suggested by the Army Corps of Engineers, designed to have minimal impact on the environment but still be able to block Katrina-strength storm surges. As shown in illustration 3, fragile I-wall barriers would be replaced with more robust T-walls, which use three rows of foundation pilings that can withstand pressure generated by hurricane-force floodwaters. A wide concrete slab, or "skirt," on the protected side deflects overflowing water that could otherwise wash away supporting soil. T-walls already installed in New Orleans held throughout Katrina without a leak.

The final takeaway: Levees and floodwalls were breached during Hurricane Katrina because poor decisions were made at varying levels of responsibility over decades, and with something as important as a city's existence at stake, design and engineering quality cannot be compromised under any circumstances.

A 17TH STREET CANAL BREACH
B LONDON AVENUE CANAL BREACHES
C INDUSTRIAL CANAL BREACH

0-2 ft.
2-4 ft.
4-6 ft.
6-8 ft.
8-15 ft.

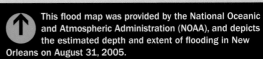
This flood map was provided by the National Oceanic and Atmospheric Administration (NOAA), and depicts the estimated depth and extent of flooding in New Orleans on August 31, 2005.

(Left): This floodwall is typical of those breached in New Orleans when water up to 25 ft. high overtopped the barriers, washing out their foundations. (Bottom left): Foundation failures were considered one possible reason for the breaches on the 17th Street and London Avenue canals. (Bottom right): This proposed T-wall barrier incorporates double foundations for stability and concrete skirts to prevent scouring of earthen levees.

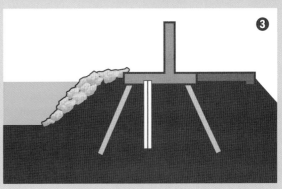

"First there was water on the street, then the sidewalk, then water in the house."

This satellite image shows Hurricane Katrina just before it made landfall as a Category 3 storm, sustaining winds of 125 mph. It did not hit New Orleans head on, instead veering slightly east, near Buras-Triumph, Louisiana. But the lack of a dead hit on the city was merely a temporary reprieve from the devastation that was soon to come to New Orleans.

and Biloxi, as well. In South Diamondhead, an entire subdivision of two hundred homes was washed away. "We rode the house," Don Haller, of Diamondhead, told the *New York Times*. Highways were flooded for miles inland, power was out across the region, and hundreds of thousands of people were displaced.

Though many thought the worst was over, New Orleans was in still more serious trouble. Katrina's 20-foot storm surge was channeled directly into the city's canals and levees. Three hours after the hurricane hit, a federal official reported that the downtown Industrial Canal had been breached. An hour later, the same was said of the Seventeenth Street Canal. Engineers suspect that in some areas the surge water topped the canals' walls, eroding the earthen foundations. In other areas, the sheer pressure of the surge appears to have destroyed the concrete walls outright. Water from Lake Pontchartrain quickly overwhelmed the pumps and poured into the city.

"We were good until the canal busted," resident Gregory Sontag said to the *Times-Picayune*, which won a Pulitzer Prize for its reporting on Katrina, even though the paper's staff was forced to evacuate its offices. "First there was water on the street, then the sidewalk, then water in the house."

By Monday night, low-lying neighborhoods such as the Lower Ninth Ward were in a state of total disaster. Daniel Weber sobbed as he told reporters about watching his wife drown as he tried to pull her onto their roof. "My hands were all cut up from breaking through the window, and I was standing on the fence. I said, 'I'll get on the roof and pull you,' and then we just went under." Weber floated on a piece of driftwood for fourteen hours before being rescued by one of the many boats plying the flooded streets searching for survivors.

Daybreak on Tuesday, August 30, brought cable

Service. "If you don't have no money, you can't go." The city established emergency shelters-of-last-resort for anyone who remained, and some ten thousand people waited out the storm in the Louisiana Superdome Sunday night.

When Katrina made landfall at 6 o'clock the next morning, the hurricane had been downgraded to a Category 4. Its eye passed east of New Orleans. But these facts hardly provided a reprieve: the storm hovered over the region for eight long hours, ripping houses apart with winds of up to 145 mph. In neighboring St. Bernard Parish, an estimated forty thousand homes were destroyed. Wind and two-story storm surges obliterated many of Mississippi's seaside communities, such as Gulfport

news footage that shocked the country. It quickly became apparent that Katrina was the most damaging hurricane in U.S. history. Eighty percent of New Orleans was flooded. News reports of looting, robbery, rape, and sniper fire in the wake of the disaster whipped media coverage into a fever pitch. But as *Popular Mechanics* reported in its March 2006 cover story, "Now What? The Lessons of Katrina," the reports of violence were grossly overplayed, and most looters were simply trying to find food and water.

By Friday the conditions at the Superdome, where the crowd had grown to twenty thousand people, forced many out into the misery of Louisiana's infamously stifling heat and humidity. In the meantime, food and water finally arrived at the New Orleans Convention Center, to which another thirty thousand people had fled. Reporters from the *Times-Picayune* interviewed people who had waded through chin-high water, and who told of friends and families vanishing in the flood. Those who remained in the city felt abandoned.

"We're being treated like animals," Donyell Porter, 25, said to a *Times-Picayune* reporter. "Look around. Man, look at the bodies. And there's no way for us to leave…. It's not right. We're humans, too."

Most of the hurricane survivors who wanted to leave New Orleans were finally able to get buses out of town on Saturday, September 3. The city they left behind had been nearly wiped off the map, but Hurricane Katrina affected 90,000 square miles in Louisiana, Mississippi, and Alabama as well. Well over 1,300 people were killed across the region, and bodies were still turning up in condemned homes eight months later. The financial tally as of July 2007 was approaching $200 billion; some predict it will top $300 billion after all the checks are written.

United States Navy sailors assigned to the dock landing ship U.S.S. Tortuga (LSD-46) search for survivors in the flooded neighborhoods of New Orleans following Hurricane Katrina. Nearly 2,000 people were killed in the hurricane and its subsequent floods, making it the deadliest hurricane in the U.S. since the 1928 Okeechobee hurricane.

The Aftermath

In the days that followed Hurricane Katrina, the wave of sentiment coursing through America was that government had failed across the board. At the federal, state, and local levels, there had been a nearly complete breakdown in communications. Despite the early-warning phone call from National Hurricane Center director Max Mayfield, nobody seems to have been prepared for—or aware of—the real extent of the damage. This seemed especially true at the top, where the director of the Federal Emergency Management Agency reported directly to officials at the White House, cutting his superior at the Department of Homeland Security out of the loop.

A bipartisan Senate report, issued in April 2006, was not hopeful about upcoming hurricane seasons. "Hurricane Katrina: A Nation Still Unprepared" cites a myriad of problems at Federal Emergency Management Agency alone, including a lack of emergency management experience at top levels, years of crippling budget shortfalls, and a deficit of experienced frontline personnel to assist in emergencies.

The rescue effort that mobilized in spite of these failings was nonetheless impressive. More than one hundred thousand emergency personnel were on the scene within three days—many within hours of Katrina's passing. More than fifty thousand people were rescued by the Coast Guard and the National Guard, not to mention ordinary citizens who piloted their own boats on impromptu rescue missions throughout the week.

Hurricanes are a fact of life for people living near equatorial waters. Sometimes, with luck, they stay out at sea and leave the mainland alone. But when they choose to make landfall, there is no stopping them. Hopefully, Hurricane Katrina will act as a wake-up call for a generation.

Approximately 80% of New Orleans was flooded in Katrina's wake. This aerial view was taken on September 11, 2005, almost two weeks after landfall. Most of the roads leading into and out of New Orleans were damaged after the storm, and the 1-10 Twin Span Bridge traveling eastbound collapsed.

SURVIVAL TIPS: EVACUATION

In the event of a hurricane evacuation, be prepared to hunker down for one week without utilities or access to grocery or hardware stores. Besides the supplies necessary for basic survival, have tools for repairing the highest-priority home damage.

The following items are essential for an evacuation—and for cleaning up after returning from one. Three days of water for four people weighs close to 100 pounds, so line up a rolling ice chest, backpacks, or another method to carry the load.

— Bottled water (1 gallon per person per day)

— Nonperishable food

— First-aid kit

— Battery-powered or hand-cranked radio

— Flashlights

— Cash

— Prescription medications

— Vital documents in a waterproof bag (deeds, insurance papers, birth certificates)

— Extra batteries (including cell phone batteries)

— Dust mask

— Duct tape

— Multi-tool

— 50 feet of heavy-duty rope

— Chain saw and fuel or bow saw and ax

— Assortment of nails, screws, and bolts

— Flexible repair clamps for burst pipes

— Tarp large enough to cover half of your home's roof

— Scrap framing lumber

— Rain gear

— Work gloves

— Fire extinguisher

— Cleaning supplies

FLOODS

The images are all too familiar: trees bent by hurricane-strength winds, homes submerged in floodwaters, rescuers motoring down city streets in fishing boats. Water is the life-giving element without which survival would be impossible. The majority of the planet and the majority of our bodies are made of water. But water has a dark side, too; not ingesting enough water causes death in days, and inhaling it causes death in mere minutes. Floods cause an average $5 billion in losses and 100 fatalities yearly—and about 3,800 towns are on flood plains. If you live in a flood zone, the U.S. Geological Survey estimates you have a one-in-two chance of experiencing a flood in your lifetime.

Heavy monsoon rains in July 2010 flooded the Khyber Pakhtunkhwa, Sindh, Punjab, and Balochistan regions of Pakistan, destroying millions of homes. The United Nations estimated that more than 20 million people were injured or homeless due to the floods, which, at one point, put one-fifth of Pakistan's total land area underwater.

YELLOW RIVER FLOODS

Centuries of Misery in China

More than three thousand miles long, China's Huang He (Yellow River) is one of the world's longest rivers. But because of its elevated river bed, the river nicknamed "China's Sorrow" is also the world's deadliest, having flooded 1,593 times in the last three to four thousand years, leading to inestimable loss of human life over the centuries. The China floods of 1931, however, brought new levels of sorrow to the country, as the rising waters of the Yellow River were responsible for the greatest natural disaster ever recorded. It's believed that the flooding in central China killed between 1.5 and 4 million people.

Called "the cradle of Chinese civilization," the Yellow River has a secure place in history as being the birthplace of the northern Chinese civilizations. Its basin was the most prosperous region in the country, although those who sought to cultivate land and societies along the river often paid a heavy price, especially in its lower course where frequent floods caused unimaginable devastation.

These floods have been responsible for the river changing its main course more than a dozen times, and at least five of those changes in the last 3,000 years were large-scale ones. What makes the river so deadly is that there's a great amount of sedimentation (loess) continuously deposited at the bottom of the river, which forms natural dams building up over time. These dams cause highly unpredictable floods, exacerbated by cyclones or continuous rain, when water is unable to reach the sea.

Flooding of the Yellow River is believed to have caused the fall of the Xin dynasty in 11 AD, when the river changed course to flow south, leading to famine, epidemics, and the massive migration of peasants who eventually rebelled and overthrew Wang Mang, then Emperor of China.

In 1887, the Yellow River offered a preview of the devastation of which it was capable during flood season. Farmers along the river had to build dikes to contain rising waters, but by September, heavy rains overcame them, creating a massive flood in Henan province. An estimated 50,000 square miles were affected and it's believed that 1 to 2 million people were killed.

The summer of 1931, however, put the Yellow River on the modern world stage. Catastrophic flooding caused untold damage to crops and industry, and there were unimaginable human casualties. It was believed that more than 1 million people died from drowning alone. The trouble began when a serious drought preceded heavy snowstorms in the winter of 1930. Then, heavy rains accompanied the spring thaw in the region, followed by an extraordinary seven cyclones that July. All the pieces were in place for disaster. And when the North China Plain began to flood over those horrific four months in 1931, millions died, and the waters of the Yellow River caused the single greatest natural disaster in recorded history.

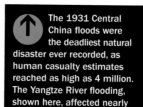

The 1931 Central China floods were the deadliest natural disaster ever recorded, as human casualty estimates reached as high as 4 million. The Yangtze River flooding, shown here, affected nearly 30 million people as rainfall in July exceeded 2 feet. Flooding along the Yangtze alone killed more than 145,000 people.

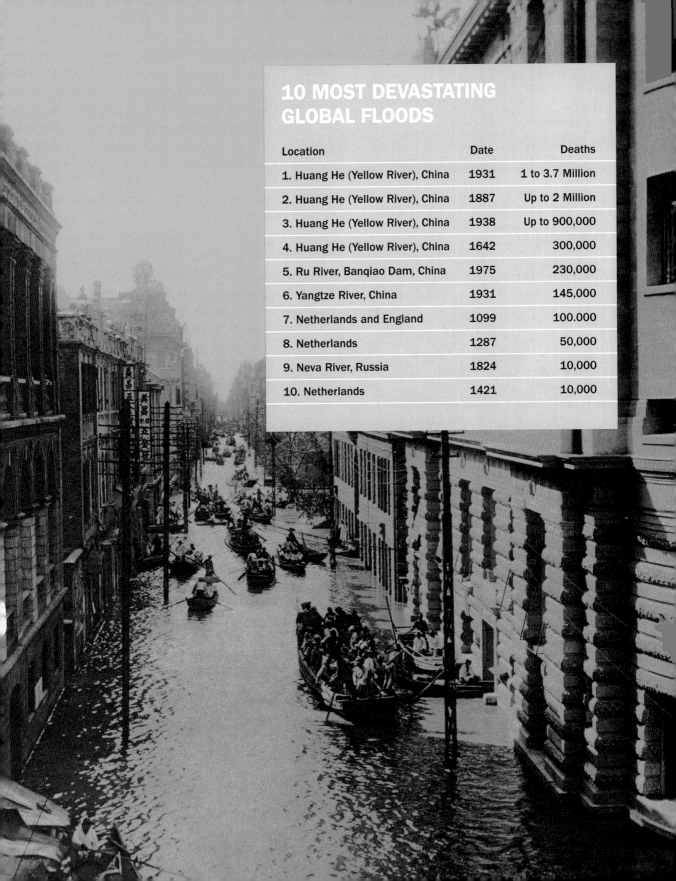

10 MOST DEVASTATING GLOBAL FLOODS

Location	Date	Deaths
1. Huang He (Yellow River), China	1931	1 to 3.7 Million
2. Huang He (Yellow River), China	1887	Up to 2 Million
3. Huang He (Yellow River), China	1938	Up to 900,000
4. Huang He (Yellow River), China	1642	300,000
5. Ru River, Banqiao Dam, China	1975	230,000
6. Yangtze River, China	1931	145,000
7. Netherlands and England	1099	100.000
8. Netherlands	1287	50,000
9. Neva River, Russia	1824	10,000
10. Netherlands	1421	10,000

In the summer of 2008, before the muck had even been cleared from neighborhoods in the heart of the Midwest, the question resurfaced: what happened? Three years after Hurricane Katrina, and fifteen years after disastrous flooding throughout the Midwest, how could the country allow rising waters to kill at least twenty-four people and cause billions in damage to structures and crops? And the Mississippi Basin isn't the only region at risk. Aging flood walls protect areas from California's Sacramento–San Joaquin River Delta to the shores of Florida's Lake Okeechobee, threatening future devastation. For many, the answer seems clear. Raise levees high enough to ensure that no flood ever wreaks havoc again.

Unfortunately, it's not so simple. For one thing, it's unclear just how high such levees should be built. "In Cedar Rapids [Iowa], nobody would have dreamed the river could get that high," says Jeffrey Schott, an instructor of urban and regional planning at the University of Iowa in Iowa City, where twenty-one buildings were flooded. The flood walls in Cedar Rapids, which had been built to a so-called one-hundred-year standard, were overwhelmed by water that rose to a five-hundred-year level. (The language is misleading: the next "five-hundred-year flood" could come next spring.)

"After '93, we could have built the levees in the Midwest 10 feet higher," says Robert Holmes, national flood specialist for the United States Geological Survey, "but the cost far exceeds the benefit." Furthermore, investing all our effort in building more and higher floodwalls can exacerbate a variety of environmental problems and create a false sense of security for home-owners and businesses that may be in the path of future flooding.

The Aftermath

Even if we could escape the law of unintended consequences, the price tag might be prohibitive. Fixing the seventy-year-old levees protecting St. Louis will cost about $200 million, according to the Army Corps of Engineers. Bolstering levees in New Orleans will cost billions, and still won't guarantee the city's safety. These investments might be worthwhile, but no one knows precisely how many more levees there are in the United States, or what it would cost to raise and strengthen them. (In 2008, the corps was asked to inventory all U.S. levees, but funding has not yet been allocated.)

If ever-higher levees aren't the answer, it may be time for a change in strategy. First, the National Flood Insurance Program, which insulates developers from much of the financial risk of building on dangerous sites, needs to be reformed. After the 1993 floods, some buildings were relocated from high-risk areas, but other flood-prone areas saw new development, thanks in part to the insurance program.

Meanwhile, experts such as Holmes are focusing on a fresh strategy for coping with floods, arming themselves and their fellow citizens with information.

It may be time for a change in strategy.

Thousands of acres of farmland were under water after a levee broke near Oakville, Iowa, in June of 2008. After months of heavy precipitation, a number of rivers in the Midwestern United States overflowed their banks for weeks on end and broke through levees at many locations. Along the Mississippi River, flood waters reached near-record heights and mandatory evacuations were ordered in several sections of Iowan cities. The fix will not come cheap. The Army Corps of Engineers believes it would cost about $200 million to fix the 70 year-old levees protecting St. Louis. In New Orleans, bolstering the levees will cost billions. And there is still no comprehensive report on which levees across the country will require strengthening and at what cost.

At St. Louis University, scientists have built a 3-D model of the Upper Mississippi Basin using aerial photographs and satellite imagery. The team tracks changes, such as new strip malls, and runs climate simulations to create detailed versions of the inundation maps used by insurance companies.

The Army Corps of Engineers has voiced concern, arguing that if individuals rely too heavily on such information, they could stop heeding the warnings of emergency managers. On the other hand, research shows that "a flood forecast of four to twelve hours could reduce [property damage] by as much as 22 percent," according to Scott Morloch, chief of the Hydrologics Network Section at the U.S. Geological Survey's Indiana Water Science Center. The new system can provide detailed warnings up to five days in advance.

The system has been tested along an 11-mile stretch of Indiana's White River. Rolling it out elsewhere would mean adding stream gauges and creating precise maps that record subtle changes in elevation across a distance of just 1 to 2 feet. The cost would run into the thousands of dollars per river mile. Yet compared to the cost of building ever-higher levees, that sounds like a bargain. We may not be able to ward off every flood, but we should be able to give people the tools to decide when to pile the sandbags higher—and when to simply clear out their belongings and head for higher ground.

MONTPELIER, VERMONT

Ice Jams Lead to Terrible Thaw: March 1992

The flood that hit Montpelier, Vermont, in March 1992 had a sneaky feel to it. The water didn't come from up the Winooski River, which skirts the city's downtown; it backpooled quietly into the streets. Spring snowmelt and steady rain had swollen the river to near its banks, breaking its thick ice cover into truck-size chunks that formed a clog in a bend just below town.

Within minutes, shopkeepers noticed water sliding across their floors. Eventually, ice cakes floated along the streets. Basement windows imploded. Propane tanks tore loose and drifted around, spewing gas. In basements throughout the business district, inventory ranging from bicycles to electric guitars sat immersed in slush. A crane and excavators broke up the jam that night, but when the water finally receded it took 650 dump truck runs to clean up the mess. Sixty-two New England towns suffered from ice jams that week, but none more than Montpelier.

The Aftermath

Unfortunately, ice jams are a chronic problem. In the past twenty-five years, they have caused serious destruction in a half dozen U.S. cities and towns—from Montpelier to Salmon, Idaho—and

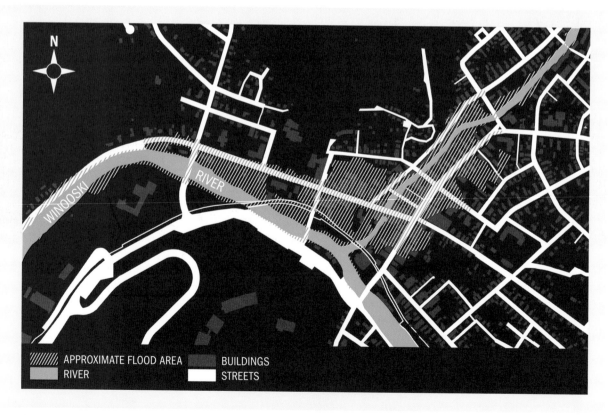

APPROXIMATE FLOOD AREA
RIVER
BUILDINGS
STREETS

WINOOSKI RIVER

An ice jam in March of 1992 brought the Winooski River in Montpelier, Vermont, to a complete stop, as water, with nowhere to go, suddenly poured over the riverbanks, flooding the heart of the small New England city. In less than an hour, more than 120 businesses were forced to shut down, and the flood caused more than $5 million in damage.

have sent smaller floods backing into towns in thirty-six states, resulting in $125 million in damage annually. This may seem minor compared to the $5 billion in damage a single hurricane can inflict, but for the tiny northern towns that ice jams most often affect, it can be calamitous.

Along with trashing businesses and homes, the flooding damages hydropower plants and, in 1996, even forced a nuclear power plant to shut down temporarily. Some high-value facilities protect themselves with dams or boom systems, but such structures are often too pricey for small municipalities, leaving residents with no real defense.

CAZENOVIA CREEK, WEST SENECA, NEW YORK ⬅

A Proactive Approach to Averting Disaster: 2006

In 2006, West Seneca, a suburb of Buffalo, New York, installed a new type of ice-control structure on Cazenovia Creek. This tributary of the Buffalo River jams every few years, causing about $1 million of damage each time. "These jams have been a nasty little problem," said town supervisor Paul T. Clark. At $2.1 million, the standard solution—a dam with piers designed by the Army Corps of Engineers—just hadn't been feasible when the town considered it back in 1987. Recently, the corps' Cold Regions Research and Engineering Laboratory (CRREL) in Hanover, New Hampshire, came up with another design.

The new pier structure consists of a line of nine concrete columns, each 10 feet tall and 5 feet in diameter, set across the creek bed at 12-foot intervals. By the time the ice melts enough to pass between the columns, it's too small to form significant jams. And the pier-only design differs in another significant way: the base of each column is flush with the creek bed. Embedded in a concrete substructure as opposed to set in a low dam, it allows fish to swim unimpeded, and stream flow and warmer-weather boat traffic to remain essentially unaltered. On Cazenovia Creek, the structure also makes use of a wooded floodplain that helps trap ice while letting water escape downstream.

"It's part dam and part valve," said CRREL's Kate White, who has been designing ice-control structures for eighteen years. "This is the new breed. Older models have weirs and gates and cost many millions. This one goes in more easily, requires very little maintenance, and is much cheaper than the previous design."

The Army Corps of Engineers is not renowned for doing things simply and cheaply. "This is a new way of thinking," White said. "We realized we don't have to completely subjugate the river. If we just hold back most of the ice, we get the same protection more economically."

Like a dam with piers, this rakelike configuration catches and holds the biggest slabs of ice, forming a jam upstream of town, where it will do no harm.

The key, White said, lies in using the correct number of columns. Build too few, and the ice escapes; build too many, and the cost climbs. To find the magic number, CRREL engineers use a computer program to create a virtual version of a particular river and model the various alternatives. They test the most promising designs on a 1:15-scale physical model in a refrigerated room at their laboratory—a reshapable "riverbed" complete with chunks of floating ice.

⬆ Because Cazenovia Creek jams with ice each winter causing nearly a million dollars in damage each time, the town of West Seneca, a suburb of Buffalo, New York, was forced to employ a fairly inexpensive pier structure designed by the Army Corps of Engineers, which traps large blocks of ice before they can float downtown and cause substantial flooding.

The pier structure on Cazenovia Creek consists of a line of nine concrete columns that each stand 10 feet tall and are 5 feet in diameter. They are embedded in concrete, and the columns line the creek at 12-foot intervals. The rakelike configuration prevents the largest chunks of ice from floating downstream, where they had been plaguing West Seneca, NY, with floods on a regular basis.

The Aftermath

At Cazenovia Creek, the full-scale pier structure took several months to install, primarily because heavy fall rains kept washing out the access road. The crew tackled half the job at a time, building temporary dams to channel the water to one side of the creek bed while working on the other, replacing the shale bottom with a heavily reinforced concrete slab, and pouring concrete into steel forms to create the row of columns.

Once set, the piers made a sturdy-looking line of soldiers. Yet during the flood in Montpelier, Vermont, in 1992, chunks of ice tore a steel railroad bridge off its abutments and turned it 90 degrees.

"An ice jam backed by a river generates a lot of force," army corps civil engineer Dustin Tellinghuisen said. CRREL designed the piers to withstand up to 450,000 pounds—a force equal to the static thrust of two 747s at full throttle, and a lot more than would be exerted by the biggest ice jams at record flows. Arm-thick, post-tensioned braided cables run integrally through each pier, anchoring them 38 feet into the bedrock.

When asked what might be strong enough to move the piers, Tellinghuisen replied, "I'm glad to say I can't think of anything."

Business owners in West Seneca look forward to seeing for themselves that the engineering project was money well spent. Adjusted for inflation, its construction cost amounted to slightly more than half that of the older design. With federal and state cost-sharing programs, the town ended up spending less than $500,000.

"That was a pretty easy sell," Clark said of the pier structure. "We've been stung enough."

WAVES OF FURY

When Winds and Quakes Cause Water Havoc

Rivers and lakes spilling over their natural or man-made barriers are not the only ways that water can wreak havoc. After Hurricane Beulah vented its fury on the Texas gulf coast in 1967, pounding wave after monstrous wave on the battered shore, survivors called it one of the worst storms of the century. Total damage from the wind and waves was estimated at $1 billion.

Yet the waves caused by this hurricane, or any hurricane, as bad as they are, are by no means the largest or most destructive. Some waves have nothing at all to do with storms.

For countless ages, people have watched waves at sea and from the shore, with reactions ranging from awe and admiration to sheer terror. Yet we have really understood very little about them.

What causes them? Where do they start? Where do they go? How do they get there? And, most importantly, how do they build up to such tremendous heights and destructive power?

The great waves that hit Galveston in 1900, since immortalized in the folk song "That Was a Mighty Day," killed seven hundred people.

In 1946, with no warning, the town of Hilo, Hawaii, was hit by a series of 50-foot waves that ripped through homes, tore up railroad tracks, demolished steel bridges, washed away beaches, and killed more than 150.

On a sunny, clear day in 1958, 30-foot crests suddenly tore out of the Atlantic Ocean onto the island of Barbados in the West Indies, hurling fishing boats onto the beaches and destroying homes along the coastline.

In 1966, while riding out a storm, the Italian liner *Michelangelo* was severely damaged by a freak wave that smashed bridge windows 81 feet above the waterline and killed three persons.

And death-dealing waves aren't always confined to open seas. In 1954, Lake Michigan suddenly sent

Winds well over 100 mph wreaked havoc in south Texas when Hurricane Beulah roared across the Gulf of Mexico in September of 1967. The Category 5 storm spawned 115 tornadoes and caused significant flooding, leading to 58 fatalities. Torrential rains in Texas (27 inches fell over 36 hours in some parts) led to an estimated $1 billion (2005 USD) in damages.

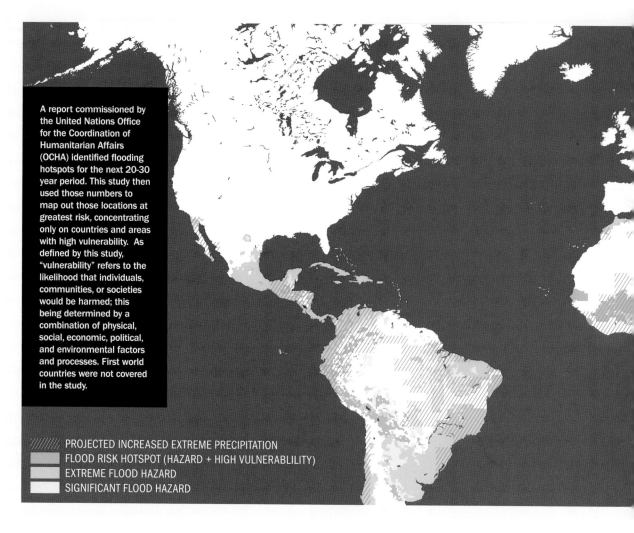

PROJECTED INCREASED EXTREME PRECIPITATION
FLOOD RISK HOTSPOT (HAZARD + HIGH VULNERABLILITY)
EXTREME FLOOD HAZARD
SIGNIFICANT FLOOD HAZARD

a giant wave roaring in on a peaceful Chicago beach, sweeping seven fishermen to their deaths.

These are just a few samples of the destructive qualities of killer waves.

Waves appear to be huge mountains of water rolling across the top of the sea. But this is just an illusion. Actually, only the wave form moves forward. The water itself merely moves up and down. It's like cracking a whip. The ripple runs down the whip, but the individual parts of the whip don't move forward at all, just up and down.

Just what sets off nature's watery whip? Many things. Wind is the main cause, and the most violent winds, of course, occur during storms. But there are other causes, including underwater explosions, such as earthquakes or volcanic eruptions, or even the topography of the body of water itself.

Let's start with wind-driven waves. Waves can be created by winds as low as 4 mph, but there are other factors involved besides velocity, including the duration of the wind and the distance over which the wind is able to act on the surface of the water. Seafarers usually refer to this distance as fetch.

In other words, a wind wave is the product of the velocity, duration, and fetch of the wind. Storms, especially storms that go on for several days, usually produce wind of the greatest velocity and longest duration, but not necessarily the greatest fetch. Storms are generally localized, but winds—

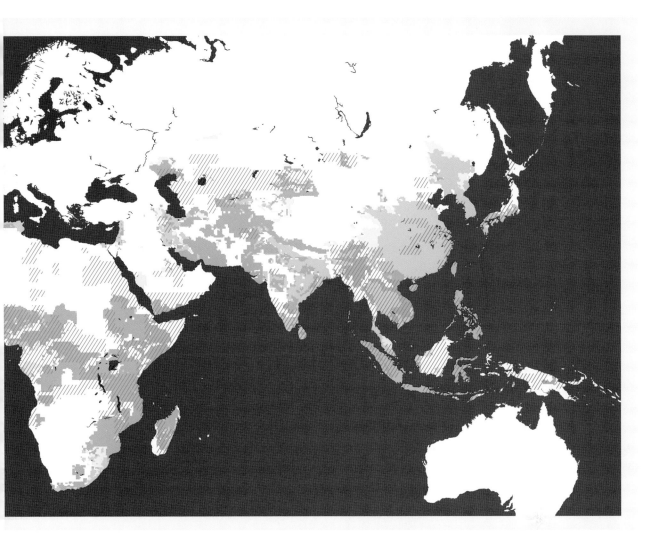

which are not necessarily associated with storms—can and do produce fetches across the greater part of an ocean. Consequently, they can produce waves of even greater destructive force than those made by local storms, including hurricanes.

The liner *Michelangelo* probably encountered such a wave. It had been running in strong winds that produced waves an average of 30 feet high from trough to crest. Suddenly, it was struck by a wave that hit its superstructure 81 feet above the waterline. Wave experts believe such a wave could have been predicted. Waves in rough seas, though not necessarily in storms, average out in height, they say—some will be higher and some lower than average. But if the rough seas and waves persist

long enough, one in twenty will be much higher than average: as much as 2.2 times higher. Such a wave probably hit the *Michelangelo*.

Waves of 60 feet are quite common in North Atlantic storms. Probably the best-measured and largest wave ever encountered, however, happened in the North Pacific in 1933. For seven days a navy oiler, the USS *Ramapo*, had been bucking weather from a storm with a fetch of thousands of miles. One night during the worst of the storm, an officer saw a great wave rising astern, its crest at a level above an iron strap on the crow's nest of the mainmast. The ship was then on an even keel, riding in a trough between waves. This made possible an exact line of sight from the bridge to

The powerful wave roared back to the Chicago shore, sweeping fishermen to their deaths.

Damage and flooding in Brownsville, Texas, after the flooding from Hurricane Beulah. An 18 foot storm surge also inundated lower Padre Island, and the force of the storm tide made dozens of cuts completely through the barrier island, causing significant damage.

the crest of the wave. Simple mathematical calculations gave the height of the wave: 112 feet!

Basically, wind waves come in two kinds: sea waves and swells. Sea waves are those produced in the immediate vicinity of a storm. They are generally a confused mass of all shapes and sizes, running into and past one another. When the wind is brisk, it blows the tops off the steeper waves, producing whitecaps.

When the waves and the storm that produced them separate, the waves are then called swells. These swells can go on in regular, undulating patterns, until they reach some distant shore and spill their foam upon a beach thousands of miles away from the original storm. Recently the Scripps

Institution of Oceanography identified individual wave groups as they originated in Antarctic storms and tracked them northward across the Pacific. About two weeks later they broke on the coast of Alaska.

If the storm that produced them was long enough and strong enough, some swells can become quite destructive when they reach shore. Almost unnoticed at sea, they can rise to great heights when they roar into shallow waters.

That's what happened to Barbados in 1958. On a pleasant day, with no storm in sight, 30- and 40-foot waves began crashing on the shore. Later it was learned that the storm that produced them had occurred two days before, near Cape Hatteras, North Carolina.

Tidal waves are just what their name implies—waves caused by the tides. But the name is often given, mistakenly, to the giant killer waves known as tsunamis. True tidal waves are normally not dangerous unless they move in the right combination with a severe storm. Tsunamis, however, are the most destructive of all waves. Triggered by landslides, volcanic explosions, or underwater earthquakes, they can move with speeds up to 500 mph.

Ironically, tsunamis are relatively harmless until they reach shore. Then they build up to devastating heights. On a clear day in 1946, Hilo, Hawaii, was unexpectedly hit by a tsunami created by an earthquake near Alaska's Aleutian Islands. The entire waterfront was destroyed. Some of the waves were recorded as far away as Chile, 8,000 miles to the south.

The third major killer wave is known as a reflected wave. It has nothing to do with the open sea but is produced in inland bodies of water that, because of their shape, are susceptible to roughly the same effects you'd get by sloshing water in a bathtub.

In 1954, one such wave took Chicago, which sits on the shores of Lake Michigan, by surprise. A squall line, moving west to east with the right combination of air pressure and wind effect, produced a long, low wave that hit the lake's southeastern shore. Now on its own, and with no place else to go but back in the direction it came from, the powerful wave roared back to the Chicago shore, sweeping fishermen to their deaths.

Reflected waves are often erroneously called seiches. A seiche (pronounced *saysh*) has similar origins but causes different results because the water continues to rock back and forth from one shore to another. Seiches occur most frequently in the United States in Lake Erie which, because of its long, narrow profile, shallow depth, and prevailing winds, seems ideally suited to them.

The Aftermath

Modern warning technologies and detection equipment, including weather satellites, have done much to neutralize the destructive forces of nature by facilitating the evacuation of people from coastal areas where a killer wave is expected to strike. But all the considerable clout modern technology wields could not save two hundred thousand lives lost all along the coasts of the Indian Ocean on December 26, 2004.

An earthquake measuring 9.1 on the Richter scale off the coast of Sumatra, Indonesia—the third largest in recorded history—was the cause of the tsunami. The seabed shifted by several feet at the epicenter, causing a massive displacement of water that became the tsunami. The wave traveled outward from the epicenter at speeds of 500 mph or more before slamming into coasts all along the Indian Ocean. The wave reached heights of up to 100 feet before striking some coasts, and in some places travelled inland over a mile.

By far the hardest hit area was Indonesia, which lost about 150,000 people. Sri Lanka, India, and Thailand were also greatly affected. In total there were over 180,000 confirmed deaths as a direct result of the tsunami; some estimates put the total closer to 230,000. It was the deadliest tsunami in recorded history. As technology improves, making tsunami prediction and tracking more reliable, and sharing information with even the most remote places on earth becomes a reality, the lessons learned from past tsunamis will no doubt reduce the toll on human life.

SURVIVAL TIPS: TSUNAMI

A wall of water crashing into the shore at 500 mph is a terrifying thing, but it's not a guaranteed killer. Staying calm and keeping a few simple guidelines in mind can increase your odds of survival.

➲ After an earthquake, beware. Many earthquakes, especially quakes whose epicenters are out to sea, cause tsunamis. Surviving an earthquake doesn't mean the danger is over, and a tremor can send tidal waves hurtling into coastlines thousands of miles away.

➲ Heed official warnings, but do not depend solely upon them. After the quake in Chile in 2010, tsunami warnings were sounded around the world. Hawaii was criticized because many felt that its response focused more on not panicking tourists—which might have decreased island revenue—than on the safety of those on the islands.

➲ Get to high ground. Whether that means climbing a city's nearest hill or charging up the stairs of a hotel or office building, the goal is to be above the tsunami when it hits. If all else fails, climb a tree and hang on.

➲ When a massive earthquake measuring between 9.1 and 9.3 on the Richter scale shook the floor of the Indian Ocean on December 26, 2004, it triggered one of the deadliest natural disasters in recorded history. A series of tsunamis with waves as high as 100 feet roared over the coastlines of the Indian Ocean, killing over 230,000 people in fourteen countries. Aftershocks were set off as far away as Alaska.

PART 2

MAN'S
ERROR

AVIATION

CHAPTER

8

Flying in a jetliner is extraordinarily safe: between 2002 and 2007, there was only one fatal crash in the United States, an astounding record considering that more than thirty thousand flights take off every day. How did flying become so reliable? In part, because of accidents that triggered crucial safety improvements.

The aughts capped 100 years of powered flight, pushing the technologies introduced in the 20th century to their limits. This past decade has seen the development of the biggest passenger airplanes, the fastest, most agile and stealthiest fighters, and the joy of flight brought to the amateur pilot as never before.

Yet despite the progress made possible by research, advances in technology, and pilot training, when equipment and people fail in the sky, the results are often spectacularly tragic.

←　Wreckage from Trans World Airlines Flight 800 stands on the runway. On July 17, 1996, just 12 minutes after takeoff, it exploded and crashed into the Atlantic Ocean near East Moriches, New York. All 230 people on board the flight for Rome, Italy were killed. Initially, it was believed that a terrorist attack was the cause of the crash, but the FBI, sixteen months later, announced that they had found no evidence of a criminal act.

There was no sign of impending tragedy 7 miles above the empty expanse of the south Atlantic Ocean on May 31, 2009, as an Air France A330 passenger jet cut through the midnight darkness. The plane had taken off three hours earlier, climbing from Rio de Janeiro on a northeast heading, its navigation computers hewing to a great circle route that would take the flight 5,680 miles to Paris.

At 10:35 PM local time, one of the copilots on the flight deck radioed the Atlantico Area Control Center in Recife, Brazil, and announced that the plane had just reached a navigation waypoint called INTOL, situated 350 miles off the Brazilian coast. The waypoint lies just shy of the Intertropical Convergence Zone, a meteorological region along the equator famous for intense thunderstorms. Staff at Atlantico acknowledged the transmission and received the airplane's reply: "Air France 447, thank you."

It was the second time within the previous twelve hours that the jet, an F-GZCP, had crossed this stretch of ocean, having flown the Paris-to-Rio leg with only two hours to refuel and load passengers before departing again. Such was the lot of the four-year-old long-haul plane: a repeated cycle of flight and turnaround, as rhythmic and uneventful as the phases of the moon. But the routine was about to be broken.

After receiving AF 447's transmission, Atlantico asked for the estimated time it would take the aircraft to reach the TASIL waypoint, which lies on the boundary of the Atlantico and the Dakar oceanic control areas. At that point communication would pass from Brazil to Senegal. But AF 447 did not reply. The controller asked again. There was still no answer. The controller asked a third and fourth time, then alerted other control centers about the lapse.

An aircraft is not supposed to vanish without a trace in this day and age. The globe is crisscrossed by constant ship and air traffic. A constellation of satellites orbits overhead, and communication is nonstop. Yet for a few days in early June, it seemed that the impossible had happened. Air France 447 and the 228 people onboard were simply gone. There was no distress call or wreckage; there were no bodies.

Within hours, the French government deployed a search-and-rescue airplane near the TASIL waypoint. Yet for days nothing was found. The only clues to the plane's fate were automatic messages that the onboard maintenance computer transmitted by a data link system called the Aircraft Communications Addressing and Reporting System (ACARS). The system sends text messages via satellite to ground stations, which then forwards them on landlines to the intended destination. In just a four-minute span, the system had broadcast twenty-four reports to Air France's dispatch center in Paris, each concerning problems with subsystems onboard the aircraft.

At 11:10 PM, about thirty-five minutes after AF 447's last verbal communication, the system sent a message that the autopilot had disconnected. Seconds later, it reported that the flight control system was unable to determine the aircraft's correct speed. Subsequent messages cited a cascade of other malfunctions. At 11:14 PM, the final message reported that the airliner's cabin had either depressurized, was moving with high vertical velocity, or both.

The story the transmissions told was tantalizing but inconclusive. And the absence of clues caused concerns that reached beyond the AF 447 investigation. Was the crash a result of pilot error, an unexpected breakdown of vital equipment, or a combination of both? Without answers, there would be no way to guarantee that another airliner wouldn't suffer the same fate.

All the attention given to a crash like Air France 447's can obscure an important truth: commercial

SUBCHEFIA DE COMANDO E CONTROLE

OPERAÇÃO DE BUSCA E RESGATE

AERONAVE A-330 AIR FRANCE AF-447

Início em 01 de junho de 2009 sem previsão de térmi

Brazil's Defense Minister Nelson Jobim, left, talks to a member of his staff in front of a diagram of the crash area of Air France 447 during a news conference in Brasilia, Wednesday, June 3, 2009. The Airbus 330-200 crashed into the Atlantic Ocean two days earlier, killing all 228 people aboard. But definitive details of the last minutes of the flight remain a mystery to investigators and officials.

air travel is incredibly safe—and getting safer. In 2008, the U.S. fatality rate was fewer than one death per nearly eleven million passenger trips. This impressive record is the result of more than a century of incremental improvements that have been amassed through painstaking forensic analyses.

"We're having many fewer accidents."

The Federal Aviation Administration (FAA) is determined to cut the already minuscule airliner fatality rate in half by 2025. With this goal in mind, the agency recently developed a new approach to making safety improvements. In 2007, it began working with airlines to sift through the masses of data that planes record about their normal flight operations, looking for safety changes that could preempt accidents, instead of learning these lessons after a plane crash occurs.

Following AF 447's disappearance, other Airbus 330 operators studied their internal flight records, looking for patterns. Delta, analyzing data from Northwest Airlines flights that occurred before the two companies merged, found a dozen incidents in which at least one of an A330's airspeed indicators—4-inch-long pressure-sensing pitot tubes located on the fuselage under the cockpit—had briefly stopped working. Each time, the flights had been

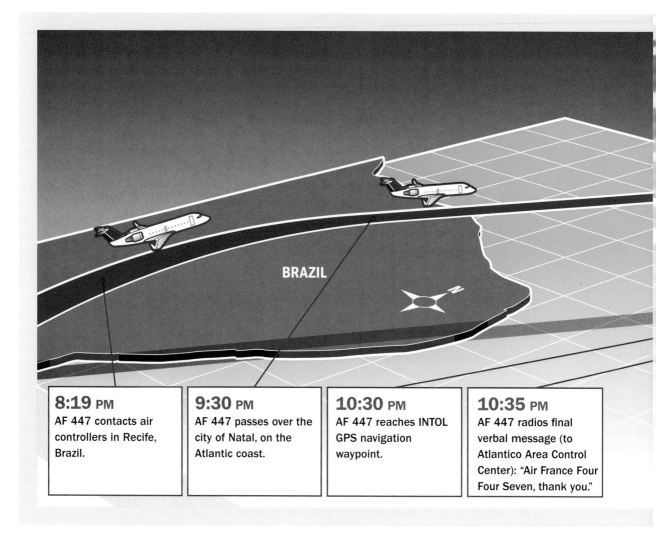

8:19 PM
AF 447 contacts air controllers in Recife, Brazil.

9:30 PM
AF 447 passes over the city of Natal, on the Atlantic coast.

10:30 PM
AF 447 reaches INTOL GPS navigation waypoint.

10:35 PM
AF 447 radios final verbal message (to Atlantico Area Control Center): "Air France Four Four Seven, thank you."

traveling through the Intertropical Convergence Zone, the same location where Air France 447 had disappeared. In the case of the Northwest A330s, the pitot tube malfunctions had been brief and harmless. But what if a severe version of the problem had struck Air France 447 amid more unforgiving circumstances?

At last, on June 6, 2009, the multinational search effort began to find evidence of the crash. The Brazilian military recovered bodies and debris floating approximately 40 miles north of the last automatic aircraft communications transmission. Over the next two weeks, search vessels retrieved fifty-one corpses from a stretch of ocean 150 miles wide, along with bits of wreckage—a section of the radome, a toilet compartment, part of a galley—that collectively added up to less than 5 percent of the aircraft. The largest single piece was the tail fin, marked with the distinctive blue and red stripes of the French national carrier.

The most important piece of the wreckage, however, remained missing. More than a month after the plane went down, and despite the joint efforts of the French and U.S. navies, its black box still hadn't been found. Given the expansive search area, the ruggedness of the undersea terrain, and the depth of the water (up to 15,000 feet), locating the recorder, let alone retrieving it, was proving to be an enormous task. Once the unit's acoustic pinger passed its thirty-day certified life span, the

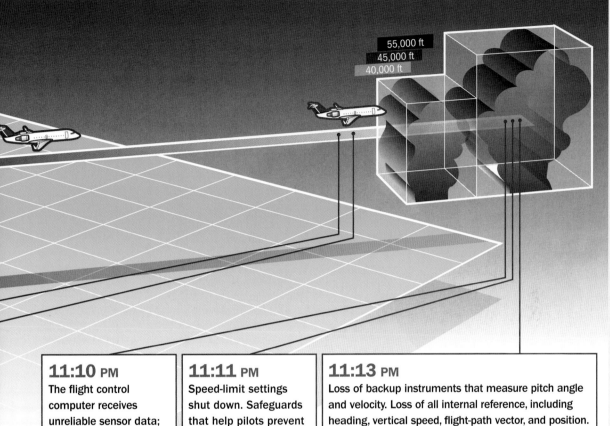

11:10 PM
The flight control computer receives unreliable sensor data; in response, autopilot disconnects.

11:11 PM
Speed-limit settings shut down. Safeguards that help pilots prevent rudder damage now fail.

11:13 PM
Loss of backup instruments that measure pitch angle and velocity. Loss of all internal reference, including heading, vertical speed, flight-path vector, and position. Last transmission: a vertical speed advisory, triggered when the cabin drops faster than 30 feet per second.

chances of recovering the black box became virtually nil.

Without the box's data, the only physical evidence of the airplane available to investigators was the mangled wreckage. From the way it had been deformed—in particular, the way the floor of the crew's rest compartment had buckled upward—French investigators determined that the fuselage hit the water more or less intact, belly first, at a high rate of vertical speed. Added to the ACARS messages and the satellite weather data, a possible scenario began to emerge.

By 10:45 PM, ten minutes after the last radio transmission, the plane hit the first small storm cell in the Intertropical Convergence Zone. Fifteen minutes later, it hit a larger, fast-growing system. And then, just before its last ACARS transmissions, the plane hit a whopper—a multicell storm whose roiling thermal energy rose more than 3 miles higher than AF 447's altitude. Buffeted by turbulence, near the heart of a strong thunderstorm, the pitot tubes froze over. Lacking reliable speed indicators, the airplane's computerized flight management system automatically disengaged the autopilot, forcing the copilots to fly the airplane manually.

Without autopilot, the pilots had no envelope protection restrictions, which are designed to keep pilots from making control inputs that could overstress the aircraft. This is particularly dangerous for airliners at high altitudes. The thin air demands

The Brazilian Air Force finds more debris from Air France Flight 447, the deadliest crash in the history of Air France. On June 6, 2009, a search and rescue operation recovered two bodies and assorted debris from the plane floating in the ocean nearly 700 miles northeast of the Fernande de Noronha islands off Brazil. Three weeks later, the search was called off after just 51 bodies had been recovered. Ultimately, officials believe that the pilots of Air France 447 flew too close to severe thunderstorms when other pilots in the air that night avoided the area completely. But because the black box was never found, investigators will likely never know for sure. Now, executives at Airbus are studying alternatives to the physical black boxes to create a system to track status of every aircraft in real time.

that airplanes fly faster to achieve lift, but they still must remain between certain speeds. Flying too fast can create a phenomenon known as mach tuck, when supersonic shock waves along the wings shift the aircraft's center of pressure aft and can make it pitch into an uncontrollable nosedive. Flying too slowly can cause a plane to stall. Disoriented in the storm, uncertain about their speed, and buffeted by turbulence, AF 447's flight crew could easily have taken the A330 outside its flight envelope.

"The fact that they didn't transmit a Mayday would seem to indicate that whatever happened to them happened quickly," said William Waldock, a professor of safety science at Embry-Riddle Aeronautical University in Prescott, Arizona.

Without more data, this kind of scenario can never be verified completely. But the global aviation community has already taken steps to prevent a similar accident. Within days, Air France replaced pitot tubes on its Airbus planes with ones made by another company, and in July 2009 Airbus advised other airlines to do the same. Three months later the FAA turned the recommendation into a regulation.

To be sure, the pitot tubes are not the definitive cause of the crash. Even if they did fail, that alone should not have been enough to bring down an airliner. As in virtually every fatal air crash, what doomed AF 447 was not a single malfunction or error in judgment, but rather a sequence of missteps that crash investigators call the accident chain.

"...[W]hatever happened to them happened quickly."

"There's always a series of events," the FAA's Jay Pardee says. "That means there are multiple opportunities to intervene and break that accident chain."

The Aftermath

In the case of Air France 447, the error chain included the copilots' decision to fly too close to severe thunderstorms—bad weather that several other pilots flying similar routes that night chose to give a wide berth. There were certainly other links in the accident chain that pushed AF 447 beyond its limits. But unless its black box is found, we may never identify those links. That means safety officials might never learn the full lessons of the disaster. To prevent a similar loss of forensic evidence, executives at Airbus say they are now studying alternatives to physical black boxes. It is feasible to create a system that could not only broadcast text messages, like ACARS, but comprehensive data about the status of every aircraft, in real time. The aircraft would continuously transmit data to VHF stations within a radius of 125 miles, or by satellite if the plane is farther away. Airliners in flight could one day stream all sorts of high-speed data, sharing information directly with one another.

"It would be a network in the sky," says Bob Smith, chief technology officer at Honeywell, which manufactured Air France 447's ACARS. "Aircraft could not only pass information about their location and where they're headed," he says, "but whole data sets. An airliner over Seattle could send its weather radar picture to a plane inbound from Dallas. And the guy from Dallas could pass it along to five other aircraft." Military aircraft already use a similar system; it is not clear if civil aviation will adopt it.

The disquieting truth is that we don't really know precisely what happened to Air France 447, and perhaps never will. The same links in the accident chain could someday take down another unlucky airliner. If they do, improved technology might provide investigators with the data they need to make sure that the next time is the last.

THE EVOLUTION OF AIR SAFETY

Ten Accidents between 1956 and 1998

In past accidents, evidence enabled scientists, engineers, and airline safety officials to piece together the exact causes of crashes and near misses and help prevent a greater frequency of tragic occurrences. Here are eight crashes and two emergency landings that have improved your chances of landing safely each time you step on a plane.

Aloha Airlines Flight 243 at Kahului Airport on April 28, 1988, after its fuselage was ripped apart during flight. The jet was just 23 minutes into the flight and cruising at an altitude of 24,000 feet when the Boeing 737 experienced an explosive decompression, tearing the top of the plane's skin off just behind the cockpit.

1956

GRAND CANYON, ARIZONA

The TWA Super Constellation and the United DC-7 had taken off from Los Angeles only three minutes apart, both headed east. Ninety minutes later, out of contact with ground controllers and flying under see-and-avoid visual flight rules, the two aircraft were apparently maneuvering separately to give their passengers views of the Grand Canyon when the DC-7's left wing and propellers ripped into the Connie's tail. Both aircraft crashed into the canyon, killing all 128 people aboard both planes.

The accident spurred a $250 million upgrade of the air traffic control system, leading to improvements that have paid dividends for air travelers across the country for decades. (There hasn't been a collision between two airliners in the United States in forty-seven years.) The crash also triggered the creation in 1958 of the Federal Aviation Agency (now the Federal Aviation Administration) to oversee air safety.

1978

PORTLAND, OREGON

United Airlines Flight 173, a DC-8 approaching Portland, Oregon, with 181 passengers, circled near the airport for an hour as the crew tried in vain to sort out a landing gear problem. Although gently warned of the rapidly diminishing fuel supply by the flight engineer on board, the captain—later described by one investigator as "an arrogant SOB"—waited too long to begin his final approach. The DC-8 ran out of fuel and crashed in a suburb, killing ten.

In response, United revamped its cockpit training procedures around the then-new concept of cockpit resource management. Abandoning the traditional "the captain is god" airline hierarchy, cockpit resource management emphasizes teamwork and communication among the crew and has since become the industry standard.

"It's really paid off," said United Captain Alfred Haynes, who in 1989 remarkably managed to crash-land a crippled DC-10 at Sioux City, Iowa, by varying engine thrust. "Without [cockpit resource management training], it's a cinch we wouldn't have made it."

1983

CINCINNATI, OHIO

The first signs of trouble on Air Canada 797, a DC-9 flying at 33,000 feet en route from Dallas to Toronto, were the wisps of smoke wafting out of the rear lavatory. Soon, thick black smoke started to fill the cabin, and the plane began an emergency descent. Barely able to see the instrument panel because of the smoke, the pilot landed the plane in Cincinnati. But shortly after the doors and emergency exits were opened, the cabin erupted in a flash fire before everyone could get out. Of the forty-six people aboard, twenty-three perished.

The FAA subsequently mandated that aircraft lavatories be equipped with smoke detectors and automatic fire extinguishers. Within five years, all jetliners were retrofitted with fire-blocking layers on seat cushions and floor lighting to lead passengers to exits in dense smoke. Planes built after 1988 have more flame-resistant interior materials.

1985

DALLAS/FORT WORTH, TEXAS

As Delta Flight 191, a Lockheed L-1011, approached for landing at Dallas/Fort Worth International Airport, a thunderstorm lurked near the runway. Lightning flashed around the plane at 800 feet, and the jetliner encountered a microburst wind shear—a strong downdraft and abrupt shift in the wind that caused the plane to lose 54 knots of airspeed in a few seconds. Sinking rapidly, the L-1011 hit the ground about a mile short of the runway and bounced across a highway, crushing a vehicle and killing the driver. The plane then veered left and crashed into two huge airport water tanks. On board, 134 of 163 people were killed. The crash triggered a seven-year joint NASA/FAA research effort, which led directly to the on-board forward-looking radar wind-shear detectors that became standard equipment on airliners in the mid-1990s. Between 1964 and 1985, major wind shear-related accidents occurred at the rate of one per year. Since 1995, Doppler radar and other on-board detectors have made wind shear accidents a rare occurrence.

1986

LOS ANGELES, CALIFORNIA

Although the post–Grand Canyon air traffic control system did a good job of separating airliners, it failed to account for small private planes like the four-seat Piper Archer that wandered into the Los Angeles terminal control area on August 31, 1986. Undetected by ground controllers, the Piper blundered into the path of an Aeroméxico DC-9 approaching to land at Los Angeles International Airport, knocking off the DC-9's left horizontal stabilizer. Both planes plummeted into a residential neighborhood 20 miles east of the airport, killing eighteen people, including fifteen on the ground.

The FAA subsequently required small aircraft entering control areas to use transponders—electronic devices that broadcast position and altitude to controllers. Additionally, airliners were required to carry traffic collision-avoidance systems, called TCAS-II, which detect potential encounters with other transponder-equipped aircraft and advise pilots to climb or dive in response. Since then, no small plane has collided with an airliner in flight in the United States.

1988

MAUI, HAWAII

As Aloha Flight 243, a weary, nineteen-year-old Boeing 737 on a short hop from Hilo, Hawaii, to Honolulu leveled off at 24,000 feet, a large section of its fuselage blew off, leaving dozens of passengers riding in the open-air breeze. Miraculously, the rest of the plane held together long enough for the pilots to land it safely. Only one person, a flight attendant who was swept out of the plane, was killed.

The National Transportation Safety Board (NTSB) blamed a combination of corrosion and widespread fatigue damage, the result of repeated pressurization cycles during the plane's more than eighty-nine thousand flights. In response, the FAA began the National Aging Aircraft Research Program in 1991, which tightened inspection and maintenance requirements for high-use and high-cycle aircraft. Since Aloha Flight 243, there has been only one American fatigue-related jet accident: United Airlines Flight 232, a DC-10 flying from Denver to Chicago, which crash landed on the runway at the airport in Sioux City, Iowa, on July 19, 1989, killing 111 of its 285 passengers.

1994

1996

PITTSBURGH, PENNSYLVANIA

When USAir Flight 427 began its approach to land at Pittsburgh, the Boeing 737 suddenly rolled to the left and plunged 5,000 feet to the ground, killing all 132 on board. The plane's black box revealed that the rudder had abruptly moved to the full-left position, triggering the roll. But why? USAir blamed the plane. Boeing blamed the crew. It took nearly five years for the NTSB to conclude that a jammed valve in the rudder-control system had caused the rudder to reverse: as the pilots frantically pressed on the right rudder pedal, the rudder went left.

As a result, Boeing spent $500 million to retrofit all 2,800 of the world's most popular jetliners. And in response to conflicts between the airline and the victims' families, Congress passed the Aviation Disaster Family Assistance Act, which transferred survivor services to the NTSB.

MIAMI, FLORIDA

Although the FAA took measures against cabin fires after the 1983 Air Canada accident, it did nothing to protect passenger jet cargo compartments—despite NTSB warnings after a 1988 cargo fire during which the plane managed to land safely. It took the horrific crash of ValuJet 592 into the Everglades near Miami to finally spur the agency to action.

The fire in the DC-9 was caused by chemical oxygen generators that had been illegally packaged by SabreTech, the airline's maintenance contractor. A bump apparently set one off, and the resulting heat started a fire, which was fed by the oxygen being released. The pilots were unable to land the burning plane in time, and 110 people died. The FAA responded by mandating smoke detectors and automatic fire extinguishers in the cargo holds of all commercial airliners. It also bolstered rules against carrying hazardous cargo on aircraft.

1996

1998

LONG ISLAND, NEW YORK

It was everybody's nightmare: a plane that blew up in midair for no apparent reason. The explosion of TWA Flight 800, a Boeing 747 that had just taken off from JFK bound for Paris, killed all 230 people aboard and stirred great controversy. After painstakingly reassembling the wreckage, the NTSB dismissed the possibility of a terrorist bombing or missile attack and concluded that fumes in the plane's nearly empty center-wing fuel tank had ignited, most likely after a short circuit in a wire bundle led to a spark in the fuel-gauge sensor.

The FAA has since mandated changes to reduce sparks from faulty wiring and other sources. Boeing, meanwhile, has developed a fuel-inerting system that injects nitrogen gas into fuel tanks to reduce the chance of explosions. In 2008, Boeing started installing the system in all newly built planes. Retrofit kits for in-service Boeings are also available.

NOVA SCOTIA, CANADA

About an hour after takeoff, the pilots of Swissair's Flight 111 from New York to Geneva—a McDonnell Douglas MD-11—smelled smoke in the cockpit. Four minutes later, they began an immediate descent toward Halifax, Nova Scotia, about 65 miles away. But with the fire spreading and cockpit lights and instruments failing, the plane crashed into the Atlantic Ocean about 5 miles off the Nova Scotia coast. All 229 people aboard were killed.

Investigators traced the fire to the plane's in-flight entertainment network, whose installation led to arcing in vulnerable Kapton wires above the cockpit. The resulting fire spread rapidly along flammable Mylar fuselage insulation. The FAA ordered the Mylar insulation replaced with fire-resistant materials in about seven hundred McDonnell Douglas jets.

AIR TRAVEL IN A POST-9/11 WORLD: AMERICAN AIRLINES FLIGHT 587

New York City, November 12, 2001

Airlines are constantly reviewing and revising their operational policies and training methods, but sometimes it takes tragedy to effect change. The crash of Flight 587 is significant not only for what happened, but when it happened; in November 2001, it was the first crash in the United States after September 11, 2001.

On the morning of November 12, 2001, an American Airlines Airbus carrying 251 passengers and 9 crew members from New York's John F. Kennedy International Airport to the Dominican Republic broke up just after takeoff and plunged from the sky, killing all aboard and 5 people on the ground. It was the second deadliest accident in American aviation. Since it occurred in New York not long after 9/11, many suspected terrorism. But experts soon focused on a more prosaic culprit: the turbulent wake left behind by a preceding aircraft flying along the same route.

It was thought that the Airbus pilot, hitting rough air, overcorrected with the rudder, creating stresses that ultimately ripped the vertical stabilizer from the plane. To most, that scenario seemed dubious: could a patch of turbulence, combined with the heavy foot of the pilot on the rudder pedal, really rip the tail from a plane? According to the National Transportation Safety Board's final report, this was what brought down

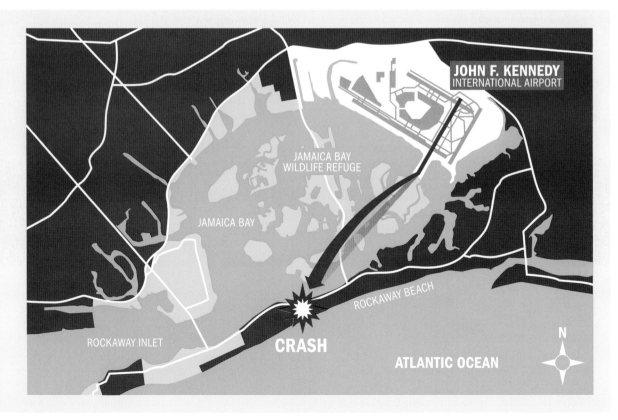

JOHN F. KENNEDY INTERNATIONAL AIRPORT

JAMAICA BAY WILDLIFE REFUGE

JAMAICA BAY

ROCKAWAY BEACH

ROCKAWAY INLET

CRASH

ATLANTIC OCEAN

N

Fires burn in New York's Belle Harbor neighborhood after the crash of American Airlines Flight 587, which killed 265 people. Tensions ran high following the crash, given that it occurred in the same city where just two months before, terrorists crashed two passenger jets into the World Trade Center. Officials concluded that a turbulent wake from a preceding plane that had been flying the same route only moments before had tragically affected Flight 587. They also concluded that the pilot over-corrected from wake turbulence, dooming the aircraft. It was the second deadliest accident in American aviation.

Flight 587. But contributing factors made the tragedy more complex—and controversial.

When a plane flies, each wingtip trails a swirling vortex that is like a horizontal tornado. If another plane enters this invisible turbulence it can suddenly rise, sink, or roll violently. Such turbulence is rarely fatal, however, because air traffic controllers maintain adequate separation between jets. The plane preceding Flight 587, Japan Airlines Flight 47, a Boeing 747, was 5 miles downrange by the time the Airbus took off—normally a sufficient distance for the turbulence to dissipate.

When Flight 587 hit the 747's vortex ninety-six seconds later, the bump was mild enough that First

"His actions were unnecessary and excessive...."

Firefighters search through debris at the scene of the American Airlines Flight 587 crash. The pilot's overreaction caused the plane to go down near John F. Kennedy International Airport in Queens, NY, just miles from the terrorist attack two months before at Ground Zero in Manhattan.

Officer Sten Molin, who was at the controls, corrected only with the ailerons. The second bump fifteen seconds later was a little worse. If Molin had done nothing, the plane would have flown through it without incident. "His actions were unnecessary and excessive, because the effect of the wake turbulence on the aircraft was minor," John O'Callaghan, an NTSB investigator, announced at the October NTSB board meeting at which the final report was reviewed. Why, then, was Molin's response so aggressive?

The Aftermath

Part of the problem lay in Molin's flying style. During the NTSB investigation, two fellow pilots said that Molin tended to overcorrect for wake turbulence. Making matters worse, American

Airlines' flight training encouraged pilots to use the rudder to recover from upsets in flight (incorrectly, Airbus maintains). Even worse, the NTSB found that the training program used a flight simulator that was unrealistic in portraying how an Airbus A300 would respond during wake turbulence. After the accident, American Airlines changed its training program.

Some critics contended that the advanced composite materials used to reduce the Airbus's weight have not stood the test of time. But the NTSB found that as Flight 587 fishtailed through the sky, its vertical stabilizer withstood the fast-building stresses as it had been designed to. The rudder was certified to handle only a full deflection—that is, to travel from neutral all the way right or left, but not to swing alternately from far right to far left. It wasn't until the rudder had done this three times, with load forces almost twice the fin's design capacity, that the vertical stabilizer finally gave way.

It is not as if Airbus was faultless, however. Compared with other commercial jets, the NTSB found that the A300 had "the lightest pedal forces of all the transport-category aircraft evaluated." In other words, it had a hair trigger, a tendency of which the pilot was fatally unaware. That a fully laden commercial jet could meet such a catastrophic fate merely as a result of pressing the rudder pedals was one that, tragically, probably surprised the two pilots as much as it did horrified observers on the ground.

The airline industry is constantly evolving; equipment and training advances make flying safer than ever before. But operating an aircraft is extraordinarily difficult even in the best conditions. The latest safety equipment and most rigorous training is still no match for simple recklessness and human error.

THE HINDENBURG EXPLOSION:

**Blaze in the Skies
on May 6, 1937**

The Hindenburg arrived at Lakehurst, NJ, on May 6, 1937, following the airship's trans-Atlantic crossing. As it was preparing to dock, it suddenly burst into flames. In just 30 seconds, the Hindenburg was gone. Ninety-seven people were aboard the aircraft and only 62 survived. The explosion's cause remains a mystery to this day.

For three and a half minutes on the night of October 14, 2004, Captain Jesse Rhodes and First Officer Peter Cesarz were on top of the world. The two Pinnacle Airlines pilots had pushed their twin-engine, fifty-seat regional jet to its maximum altitude and were now proud members of the "410 Club," an unofficial society of Pinnacle airmen who have attained 41,000 feet in a Bombardier CRJ200.

Rhodes went to the galley to grab cold Pepsis to celebrate. He was barely settled back in the cockpit when the plane's radio crackled. "Are you an RJ [regional jet] 200?" inquired an air traffic controller in Olathe, Kansas. "I've never seen you guys up at forty-one there."

"Yeah…. We don't have any passengers on board so we decided to have a little fun and come on up here," Rhodes replied.

The fun was short-lived. Moments later, both engines flamed out.

The fate of Flight 3701 was the subject of an intensive investigation by the NTSB, which issued its final report on the accident in January 2007. The report focused on "flight crew training in the areas of high-altitude climbs, stall recognition and recovery, and double-engine failures, [as well as] flight crew professionalism…." The report, the pilots' taped cockpit conversations, and hearings in June 2005 suggested a scenario involving poor judgment, insufficient training, and the complications that can occur when a plane is pushed beyond its capabilities.

Rhodes, 31, and Cesarz, 23, were on what pilots call a deadhead, transferring an empty plane over-night from Little Rock, Arkansas, to Minneapolis so it could be ready for a morning flight. The Canadian-built CRJ200 is Pinnacle's workhorse, making short and midrange hops, mostly in the Midwest. From wheels up, it was clear that Rhodes and Cesarz intended to see what the CRJ200 could do.

Four seconds after takeoff, at 9:21 PM, the two pilots did a pitch up, a maneuver that pinned them in their seats with 1.8 g's of force and momentarily triggered an alert from the airplane's stall warning system.

Minutes later, Rhodes and Cesarz again yanked back on the control column, rocketing the plane upward and generating over 2.3 g's of force before they eased off the flight controls. After briefly leveling off at 37,000 feet, the crew set the autopilot to climb at 500 feet per minute—more than twice the fastest recommended rate—to the airplane's maximum altitude of 41,000 feet.

As the plane rose, it succumbed to the physics of high-altitude flight: thin atmosphere offers less lift and robs the engine of air. Stuck in an aggressive climbing mode, Flight 3701's speed began to drop. To maintain the rate of climb, the autopilot angled the nose of the aircraft up, slowing it further. By the time the aircraft reached 41,000 feet and leveled off, it was flying slowly, at 150 knots indicated airspeed, and was perilously close to losing aerodynamic lift—or stalling.

"This thing ain't gonna [expletive] hold altitude," Cesarz said.

"It can't man," Rhodes replied. "We [cruised/greased] up here, but it won't stay."

The combination of high altitude and low speed once again triggered the Bombardier's stall warning system. First, "stick shakers" rattled the control columns and disengaged the autopilot to alert the crew of an imminent stall. When the crew didn't lower the plane's nose to gain speed, "stick pushers" forced the control columns forward. The flight data recorder shows that Rhodes and Cesarz overrode the stick pushers three times and forced the plane's nose back up. At 9:55 PM, as they pulled up for the last time, both engines flamed out.

"We don't have any engines," one of the pilots said.

Pinnacle Airlines Flight 3701 glided to within sight of the Jefferson City, Missouri, airport, then crashed behind a row of houses. An investigation concluded that Captain Jesse Rhodes and First Officer Peter Cesarz, piloting an empty plane, pushed their twin-engine jet to maximum altitude, only to suffer a fatal flame out minutes later. The report also stated that the pilots made a series of poor choices in attempting to restart the engines and make an emergency landing. In addition, investigators concluded that the pilots overrode the autopilot's safety system in order to push the envelope on the aircraft's capabilities at high altitude. The loss of aerodynamic lift was too much for the pilots to overcome.

While the altimeter spun downward, the crew hurriedly reviewed their options for restarting the engines. At that altitude, there were six suitable airports within reach for a forced landing. Despite the serious nature of their predicament, the pilots did not notify air traffic control of their situation or request emergency landing clearance.

First, they tried a "windmill restart" by diving to increase airspeed. The maneuver is intended to force air into the engine housing, spinning the rotors and creating enough compression for ignition. The procedure requires at least 300 knots of airspeed. But at 20,000 feet and only 236 knots, the crew pulled up and decided instead to try a second option: drop to 13,000 feet and attempt to relight the engines using the plane's auxiliary power unit, which generates pneumatic pressure to spin the engine's core.

Rhodes and Cesarz tried at least four times to jumpstart the engines using the auxiliary power unit; on each attempt, the engine cores recorded no rotation.

At 10:03 PM, the crew radioed air traffic control that they had a single-engine failure. Five minutes later, at an altitude of 10,000 feet and descending at 1,500 feet per minute, Rhodes and Cesarz were running out of options for restarting the engines. Finally, twelve minutes after the twin flameout, they revealed to air traffic control that they had a double engine failure. The plane's landing choices were now limited to two airports.

"We're gonna hit houses, dude."

With the runway lights of Jefferson City, Missouri, airport in sight but altitude slipping away, Rhodes and Cesarz realized they were in big trouble.

"Dude, we're not going to make this," Rhodes said. "We're gonna hit houses, dude."

They crashed 2.5 miles shy of the runway, behind a row of houses. On impact, the plane split apart, flipped over, and caught fire. Rhodes and Cesarz were killed. Miraculously, no one on the ground was injured.

An area of contention during the NTSB hearings about Flight 3701 had been whether a condition called core lock contributed to the fatal crash. Under normal conditions, the rotating parts inside a General Electric CF34-3B1 turbofan engine slip by each other in a finely tuned choreography. However, when an engine is shut down suddenly at high torque and high altitude—and it isn't restarted immediately—metal parts inside the engine begin to cool and contract at different rates. In rare cases, metal can contact metal and prevent the core from rotating freely, resulting in core lock.

The Air Line Pilots Association suggested that core lock, rather than pilot error, might have been the primary cause of the accident. But a review of the flight data recorder made clear that the pilots made a series of poor, and potentially fatal, decisions irrespective of whether the engines experienced core lock.

The Bombardier has a 41,000-foot service ceiling. However, according to the climb profiles in the crew's flight manual, the maximum altitude for the 500-foot-per-minute climb the pilots set was only 38,700 feet, based on the atmospheric conditions and the aircraft's weight that night. By operating outside the airplane's performance envelope, Rhodes and Cesarz subjected the engines to tremendous stress. The flight data recorder showed that soon after the crew ignored the fourth stall warning, the core temperature of the right engine reached 2,200 degrees—almost 600 degrees above the temperature achieved when engines reach redline values.

When investigators pulled apart the right engine, they found that the ends of the high-pressure turbine blades had liquefied, resolidifying on the low-pressure blades behind them. This led some industry experts to question if the right engine ever could have restarted.

IN CASE OF FLAMEOUT

1. Initiate a windmill restart using airspeed gained in a rapid dive to spin crucial engine parts.
2. Attempt an auxiliary power unit restart—usually below 15,000 feet—relying on pneumatic pressure to restart the engine.
3. Prepare for a forced landing at the nearest airport.

The Aftermath

The NTSB's report concluded that "[n]o physical evidence of core lock was found inside the engines." General Electric helped the NTSB disassemble Flight 3701's engines. Edward Orear, GE's former program manager for the CF34-3B1 engine, testified to the NTSB that neither engine showed evidence of core lock.

The data recorder showed that the pilots failed to follow proper procedures for restarting a flamed-out engine when they pulled out of their dive before reaching the necessary speed to spin the core. The report determined that binding prevented the left engine from restarting when the pilots windmilled and used the auxiliary power unit.

Since the crash, Pinnacle has set a ceiling of 37,000 feet for all CRJ200 flights. It has also added detailed climb profiles to the pilots' quick-reference guides. And the airline has modified its simulator training program, incorporating dual engine failure and stall recovery scenarios.

Although Flight 3701 ended tragically, it illustrates how many safety features protect commercial passengers. The crew misused the autopilot, took the plane outside its envelope, and repeatedly overrode the safety system. One pilot familiar with the investigation noted, "It's an object lesson in how many things you have to get wrong in order to crash your plane."

In fact, the skies have become safer thanks to advances in technology, research, and training. Perhaps there is no better example than the heroic ditching of US Airways Flight 1549 on January 15, 2009, when Captain Chesley "Sully" Sullenberger safely landed his Airbus A320 on the Hudson River off the coast of Manhattan. All of the passengers and crew survived with minimal injuries. That all of these people lived, some experts say, is a testament to the sophisticated "fly-by-wire" technology—in essence, a computer flight-control system designed by French engineers. Others point to the actions of Captain Sullenberger, who calmly guided the plane to a perfect landing on the cold waters of the Hudson. What can't be disputed is the fact that both humans and technology were aided by research and knowledge gained from past failures and horrific tragedies in the skies.

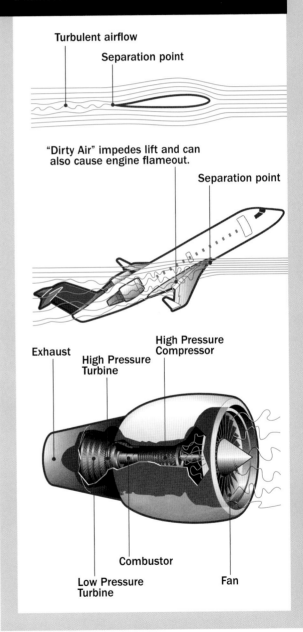

ANATOMY OF A FLAMEOUT

Air usually flows smoothly over a wing. If a plane's airspeed is too low or the nose is angled too steeply upward, the air becomes turbulent. This "dirty air" reduces lift, ultimately leading to a wing stall. It also can interfere with airflow into the engines and extinguish combustion—a condition known as a flameout.

Turbulent airflow

Separation point

"Dirty Air" impedes lift and can also cause engine flameout.

Separation point

High Pressure Compressor

Exhaust

High Pressure Turbine

Combustor

Low Pressure Turbine

Fan

SURVIVAL TIPS: FREE FALL

You're alone, six miles up, and falling without a parachute. Though the odds are long, a small number of people who found themselves in similar situations lived to tell the tale.

○ Oxygen is scarce at 35,000 feet. By now, hypoxia is starting to set in. You'll be unconscious soon, and you'll cannonball at least a mile before waking up again.

○ At 22,000 feet, you've got roughly 2 minutes until impact. To slow your descent, emulate a sky diver. Spread your arms and legs, present your chest to the ground, and arch your back and head upward. This adds friction and helps you maneuver.

○ Studies of bridge-jump survivors indicate that a feet-first, knife-like entry (aka "the pencil") best optimizes your odds of resurfacing in a water landing. For hard surfaces, some sources recommend a wide-body impact, but others believe the classic sky diver's landing stance—feet together, heels up, flexed knees and hips—best increases survivability.

○ If you somehow manage to survive, it's not over yet. "Your best bet is always staying with the plane," says Peter Kummerfeldt, former director of survival training at the U.S. Air Force Academy. "It's just easier to find a plane than a person." Plus, that plane, if it isn't too badly damaged, has resources: shelter, fuel for fires, and foam rubber for insulation.

↑ US Airways Flight 1549 was headed from New York to Charlotte, North Carolina, on January 15, 2009, when it struck a flock of Canada Geese shortly after takeoff from LaGuardia Airport. The bird strike led to an immediate loss of thrust from both engines and the crew quickly determined that they would not be able to reach an airfield. Instead, the plane was ditched in the Hudson River where it remained only partially submerged until rescue crews could arrive. There was no loss of life.

Astronaut Bruce McCandless II, mission specialist, participates in an extra-vehicular activity a few meters away from the cabin of the shuttle Challenger during the STS-41B mission in April 1983. He is using a nitrogen-propelled hand-controlled Manned Maneuvering Unit. This was the first time an astronaut performed a spacewalk without being tethered to the shuttle.

SPACE

The moon landing of July 20, 1969, was the culmination of thousands of years of human fascination with all things beyond the earth's atmosphere. Yet the act of ascending to the stars is complex: spacecraft contain many moving parts, and missions require detailed planning and extremely precise calculations. A misplaced decimal point could skew the course by hundreds of miles, and the more people and technology involved, the greater the chance for tragedy.

During the height of the shuttle program in the 1990s, six or seven shuttles and about 40 astronauts per year were launched into orbit for science and defense purposes. NASA's budget has been repeatedly cut since then, and now, starting in 2011 and for the foreseeable future, just four Americans will make it into space annually—as passengers on foreign rockets. In more positive terms, NASA's focus has shifted to the much more challenging issue of how to get people beyond earth's orbit.

Private industry is poised to pick up the slack, as they have done for the delivery of multi-hundred-million-dollar satellites for a couple of decades. While it's not the end of human spaceflight, NASA is abandoning the field of getting humans to orbit, and is handing the ball to the new players.

"Hey, we've got a problem here."

It was this laconic report from the crew of Apollo 13 on April 13, 1970, to Mission Control in Houston that let the world know that the flight was in trouble. Some 200,000 miles from earth as it hurtled toward the moon, something went horribly wrong. A liquid-oxygen fuel tank exploded, crippling the spacecraft. The three astronauts were suddenly faced with a grim possibility: they could become the first men stranded in space. That they ultimately avoided this fate was a tribute to their training and composure, and to the engineering know-how available to them via radio from Houston.

As a waiting world was to learn, smooth coordination between the astronauts and Mission Control solved the problem. Using power and life-support systems of the lunar module intended for the moon landing, Apollo 13 swung around the moon and returned to the vicinity of the earth. The astronauts conserved food and water for the long trip back, but each lost significant amounts of weight and became dehydrated.

The explosion also damaged the alignment optical telescope as well as the onboard computer, making it impossible for the crew to sight stars in order to determine the correct trajectory back toward earth. NASA quickly developed an alternative procedure, enabling the astronauts to use the sun as an aligning star, and the crew could hardly contain their excitement when they discovered that this calculation (which proved to be less than a half degree off) had been obtained. On Apr. 17, 1970, the command module carrying Jim Lovell, Tom Mattingly, and Fred Haise made a safe reentry.

The Aftermath

In Houston, relief was mixed with the knowledge that a grim lesson had been learned. Robert R. Gilruth, then director of NASA's Manned Spacecraft Center (now the Johnson Space Center), put it this way: "I think it has been made quite clear, more than by any of the Apollo flights up to this point, that flying to the moon is not just a bus ride."

The fate of the space shuttle Challenger shows just how close each mission can be to disaster. On January 28, 1986, the Challenger exploded seventy-three seconds into its tenth flight, killing all seven astronauts on board. The disaster started with a simple faulty O-ring gasket on one of the rocket boosters, which led to an explosion in the fuel tank.

It is telling to look back at the capabilities of NASA scientists and engineers in 1981. There hadn't been a manned space launch since 1975, and computers were nowhere near as powerful or sophisticated as they are today. Still, many of these same scientists had been pioneers of space technology and had been involved when man first walked on the moon. But the space shuttle program presented many challenges that the Apollo missions had not.

"...[F]lying to the moon is not just a bus ride."

Crewmen aboard the U.S.S. Iwo Jima, prime recovery ship for the Apollo 13 mission, hoist the Command Module aboard the ship. The Apollo 13 spacecraft splashed down at 12:07:44 PM, April 17, 1970, in the South Pacific Ocean. Despite the mission's troubles, the Command Module splashed down just 4 miles from the recovery ship and only seconds off schedule. The astronauts inside the capsule were Commander James A. Lovell, Command Module Pilot John L. Swigert, and Lunar Module Pilot Fred W. Haise. Swigert was a late replacement for Command Module Pilot Ken Mattingly who was grounded due to exposure to German measles.

STRANDED BEYOND THE SKY:

When Survival Is Success

An astronaut stranded in a crippled craft only 200 miles from Earth is as vulnerable as if he were adrift 200,000 miles away. But for the quick thinking and ready communication of the Houston team, the Apollo 13 mission would have had a greatly different foot-note in history.

Since the "successful failure" of Apollo 13, NASA has spent untold billions in an attempt to make things safe over the decades—and still seventeen astronauts have died in the process. Today, NASA's official overall probability risk assessment number (PRAN) for complete loss of life and vehicle for the Space Shuttle remains at the widely quoted 1/100. Theoretically, there's a one percent chance of catastrophe. While the agency stresses that Space Shuttle flights are extremely dangerous missions—sending humans into space always is—officials worry that the risk has been overstated. According to NASA, in practice, it's not as bad as it sounds.

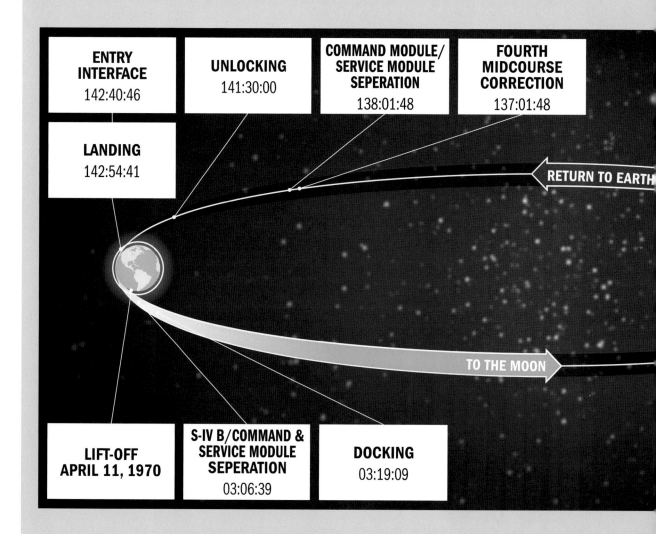

ENTRY INTERFACE
142:40:46

UNLOCKING
141:30:00

COMMAND MODULE/ SERVICE MODULE SEPERATION
138:01:48

FOURTH MIDCOURSE CORRECTION
137:01:48

LANDING
142:54:41

RETURN TO EARTH

TO THE MOON

LIFT-OFF APRIL 11, 1970

S-IV B/COMMAND & SERVICE MODULE SEPERATION
03:06:39

DOCKING
03:19:09

"This one in one-hundred is the probability risk assessment number given," NASA space operations spokesman Allard Beutel said. "It's the chance of the possibility of losing the space shuttle and crew." It's not, however, the odds of losing the crew on any given mission. "It doesn't mean that if you have a hundred launches, you're going to have an accident," says Beutel.

For these olds to make sense, Karl Sigman, a Columbia University engineering professor, explained, it must be considered over a large number of tries. The larger the number flips, the more accurate the coin flip probability becomes. According to Sigman, the 1/100 chance of a catastrophe would only apply after a large number of launches. It's called the Law of Large Numbers: the more frequent an event, the more likely you are to see the pattern. So what, exactly, are the odds of catastrophe? Sigman calculates that with a 1 in 100 PRAN for a complete loss of life and vehicle, it's now safer than climbing Mt. Everest.

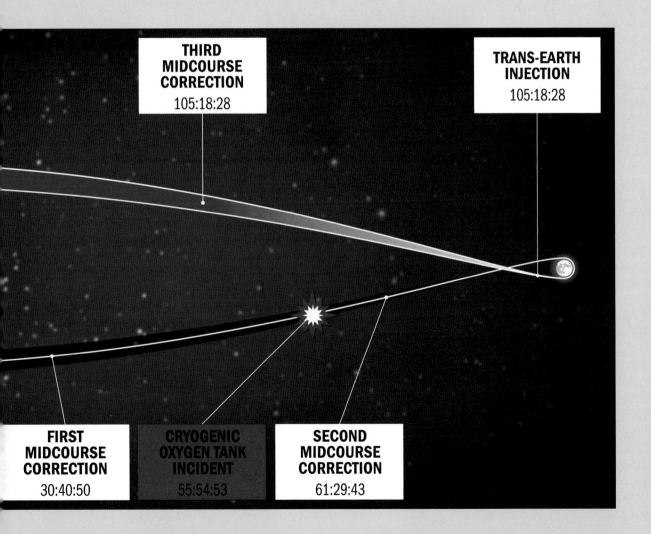

THIRD MIDCOURSE CORRECTION
105:18:28

TRANS-EARTH INJECTION
105:18:28

FIRST MIDCOURSE CORRECTION
30:40:50

CRYOGENIC OXYGEN TANK INCIDENT
55:54:53

SECOND MIDCOURSE CORRECTION
61:29:43

CHALLENGER AND COLUMBIA

Space Shuttle Missions End in Tragedy

Most middle-aged adults can still tell you where they were the moment they learned that the Space Shuttle Challenger was destroyed not even two minutes into the mission, killing all seven crew members on January 28, 1986.

Yet despite the fact that space shuttle missions had become so common and error-free that they were barely covered by the media, most people were shocked when on February 1, 2003, the Space Shuttle Columbia was destroyed during re-entry, killing all seven astronauts on board.

With NASA set to retire the Space Shuttle program in 2011, there are undoubtedly many successes and accomplishments to look back on. The Hubble Space Telescope was launched and repaired by the shuttle fleet, and the program participated in the International Space Station for years. The Space Shuttle's lasting legacy may well be that it made space travel routine in the eyes of the world. But much was learned from the tragedies of the Challenger and the Columbia disasters. Here's a look back at what went wrong in both missions:

Challenger

Just 73 seconds into its flight, the Space Shuttle Challenger broke apart as television cameras capture the dramatic and tragic last moments. The resulting investigation was unable to determine the exact timing of the crew's death, though it is believed that at least a few of them survived the initial breakup of the spacecraft. Because the shuttle was not built with an escape system, none of the surviving astronauts were able to survive after the crew compartment crashed into the ocean.

The tragedy put the shuttle program on a nearly three-year hiatus as the Rogers Commission (appointed by President Ronald Reagan)

investigated the accident. The Commission concluded that the rubber O-rings that sealed the joints of the solid rocket booster were faulty, allowing pressurized gas to reach the external fuel tank, causing the tank to structurally fail.

The Commission also found that NASA managers had known about the potential problem with the O-rings and had disregarded warnings from engineers that launching the shuttle in cooler temperatures could have dangerous consequences.

CHALLENGER AND COLUMBIA: BY THE NUMBERS

	CHALLENGER	COLUMBIA
First Flight	April 4, 1983	April 12, 1981
Prior Missions	9	27
Number of Orbits	995	4,808
Time in Space	62 days	300 days
Distance Traveled	25,803,939 miles	125,204,911 miles
Satellites Deployed	10	8

Unusually low temperatures just prior to liftoff also figured into the Challenger's destruction. The O-rings were supposed to seal the joints between segments of the solid-fuel rocket boosters, but the failure allowed a plume of flame to leak. Ultimately, a rapid release of hydrogen combined with oxygen leaking from the intertank led to a rupture of Challenger's reaction control system and the stress caused the Challenger to break apart. The disaster led to NASA's grounding of the shuttle fleet for two-and-a-half years.

Columbia

The Space Shuttle Columbia had 27 successful missions into space, but it was the 28th mission on January 16, 2003, that is remembered most. After a 16-day scientific mission, Columbia disintegrated while re-entering the earth's atmosphere. The Columbia Accident Investigation Board examined nearly 84,000 pieces of debris that were collected from Texas and parts of Louisiana.

In their investigation, the Board found that Columbia was damaged during the launch two weeks earlier, when a piece of foam insulation no bigger than a breadbox broke off the shuttle's main propellant tank under the force of liftoff. The debris crashed into the leading edge of the left wing, causing fatal damage to the thermal protection system that protects the shuttle from the intense heat generated on re-entry.

Some NASA engineers suspected that the wing had been damaged, but managers were reluctant to begin a more complete investigation while Columbia was still in orbit, as it was generally conceded that mid-mission repairs would have been impossible.

The Space Shuttle Columbia had 27 successful flights before its fateful explosion in 2003. Pictured here is its October 22, 1992, launch, a ten-day mission with a crew of five NASA astronauts and a Canadian Payload Specialist. The photograph was taken by astronaut Steven R. Nagel from a Shuttle Training Aircraft.

SPACE THREAT

January 11, 2007 and Beyond

When the Berlin Wall fell, people all over the world breathed a collective sigh of relief. The worldwide arms race was over…for a time. But on January 11, 2007, China committed an act that may have sparked a new arms race, one that will be run in a surprising new arena: space.

At 5:28 PM EST, a satellite arced over southern China. It was small—just 6 feet long: a tiny object in the heavens, steadily bleeping its location to ground stations below, just as it had every day for the past seven years. And then it was gone, transformed into a cloud of debris hurtling along at nearly 16,000 mph along the main thoroughfare used by orbiting spacecraft.

It was not the start of the world's first war in space, but it could have been. It was just a test: the satellite was a defunct Chinese weather spacecraft. And the country that destroyed it was China. According to reports, a mobile launcher at the Songlin test facility near Xichang, in Sichuan Province, lofted a multistage solid-fuel missile topped with a kinetic kill vehicle. Traveling nearly 18,000 mph, the kill vehicle intercepted the sat and, boom!—obliterated it.

For China, a nation that has already sent humans into space and developed intercontinental ballistic missiles (ICBMs), the technology involved in the test was hardly remarkable. But as a demonstration of a rising military posture, it was a surprisingly aggressive act, especially since China has long pushed for an international treaty banning space weapons. "The move was a dangerous step toward the abyss of weaponizing space," said Theresa Hitchens, director of the Center for Defense Information, an independent defense research group in Washington, DC. "China held the moral high ground about space, and that test reenergized the China hawks in Congress. If we're not careful, space could become the new Wild West. You don't just go and blow up things up there."

Every industrialized country relies on satellites every day, for everything from computer networking technology to telecommunications, navigation, weather prediction, television, and radio. This makes satellites especially vulnerable targets. Imagine the U.S. military suddenly without guidance for its soldiers and weapons systems, and its civilians without storm warnings or telephones.

The United States and Russia, the two countries with proven antisatellite—or ASAT—capabilities, have long steered clear of satellites as military targets. Even during the Cold War, spy sats were hands-off; the consequences of destroying them were far greater than those of unwelcome surveillance. "The consensus," Clark said, "was that anybody could look at anybody else."

Nevertheless, the U.S. military has spent decades designing weapons capable of taking out other countries' satellites. The crudest American ASAT test, code-named Starfish Prime, took place in 1962, when the U.S. Air Force detonated a 1.4-megaton nuclear weapon at an altitude of 250 miles. The explosion, which occurred about 800 miles west of Hawaii, disabled at least six U.S. and foreign satellites—roughly one-third of the world's low earth orbit total. The resulting electromagnetic pulse knocked out 300 streetlights in Oahu. Clearly, nukes worked as ASAT weapons—but far too indiscriminately.

To develop a more surgical capability, the air force launched Project Mudflap, which was designed to destroy individual Soviet satellites with missiles. But inaccurate space-guidance systems plagued early tests. Then, on May 23, 1963, the air force pulled off a successful intercept with a modified Nike-Zeus ballistic missile launched from Kwajalein Atoll in the Marshall Islands. It took out

"You don't just go and blow up things up there."

The free-flying Tracking and Data Relay Satellite-E (TDRS-E), still attached to an Inertial Upper Stage (IUS), was photographed by one of the crewmembers during the STS-43 mission. The U.S. Military has spent decades designing weapons capable of taking out other countries' satellites. This, despite the fact that even during the Cold War, the United States and Russia had something of an unwritten agreement to steer clear of shooting down each other's satellites. The consensus at the time was that such an aggressive act by either country could spur an incident that simply wasn't worth whatever surveillance might be gained.

The Explorer XVII Satellite: Weighing 405 lbs., this 35-inch pressurized stainless steel sphere measured the density, composition, pressure and temperature of Earth's atmosphere after its launch from Cape Canaveral on April 3, 1963. The mission was one of three that Goddard Space Flight Center specifically conducted to learn more about the atmosphere's physical properties— knowledge that they ultimately used for scientific and meteorological purposes. Not long after, the United States developed far more sophisticated technologies both in satellites and the air-launched missiles that were designed to bring them down.

a rendezvous and docking target for NASA's Gemini missions at an altitude of 150 miles.

Over the next several decades, the U.S. Air Force graduated to more sophisticated air-launched missiles that could hit their targets with far better accuracy. In 1985 the United States destroyed an American solar observation satellite using a three-stage, heat-seeking miniature vehicle fired from an F-15 fighter jet. That test, like the Chinese one in 2007, used a kinetic kill vehicle that spewed debris into space. Funding for the program was cancelled, however, before the air-launched system could be perfected.

That same year, at the White Sands Missile Range in New Mexico, the air force began operating the

"It's an act of war."

powerful mid-infrared advanced chemical laser, or MIRACL. In 1997, it was used to temporarily blind sensors on an air force missile-launch and tracking satellite. The sat remained intact; no debris was created.. Since 2007, the Starfire Optical Range at Kirtland Air Force Base in New Mexico has fired lasers at low earth orbit satellites. Some $400 million has been spent in recent years to develop another sophisticated kill vehicle: a three-stage missile that smacks an enemy's craft with a sheet of Mylar plastic, disabling it without producing any debris. It has yet to be fully tested, and would only work on satellites in low earth orbit; communication and GPS sats are too high.

Destroying an adversary's satellites has far-reaching implications. A civilian weather satellite used for tracking hurricanes could also watch military movements. Many satellites are used by multiple nations. And once a nation disables an adversary's satellites, it puts its own in peril. As Charles Vick, a senior analyst at GlobalSecurity, a public policy organization based in Alexandria, Virginia said, "It's an act of war."

So why did China risk provoking international hostility? The country's government has been opaque. "The experiment is not targeted at any other country," said a foreign ministry spokeswoman in Beijing.

Some experts think at least part of China's motivation lies in an unclassified 2006 U.S. report on the future of military activities in space. The document reaffirms that "[t]he United States considers space capabilities…vital to its national interests. Consistent with this policy, the United States will: preserve its rights, capabilities, and freedom of action in space; dissuade or deter others from either impeding those rights or developing capabilities intended to do so…and deny, if necessary, adversaries the use of space capabilities hostile to U.S. national interests."

BATTLEFIELD SPACE

Publicly available satellite images of the Songlin test facility, near Xichang, Sichuan Province, were censored by China prior to the test; immediately following, they were declassified.

Experts believe a four-stage, solid-fuel launch vehicle—thought by American officials to be based on China's KT-2—boosted the kill vehicle into orbit.

The Feng-Yun 1C target vehicle was launched May 10, 1999, with an active life span of two years. The 1650-pound polar-orbiting weather satellite was equipped with two, ten-band scanning radiometers for earth observation.

The kinetic-energy kill vehicle was traveling nearly 18,000 mph when it struck the satellite.

The collision blasted debris throughout low earth orbit. Within seven hours, a band of debris had spread halfway around the globe.

China's antisatellite strike took place at an altitude of 537 miles, which places the international spacecrafts depicted below well within range of China's space arsenal.

Some space junk, however, doesn't come from ASAT missiles. Some flotsam, such as spent boosters, comes with the territory—and others result from indifference or lousy planning, such as the bags of garbage Soviet cosmonauts jettisoned from the MIR space station for 15 years. As of July 2007, here's how the three biggest space players stack up.

	Satellites Launched	Debris in Orbit	Debris per Satellite
UNITED STATES	1773	3176	1.79
RUSSIA	3251	2917	0.89
CHINA	104	1822	17.52

"...[T]he Chinese test was an effort to force the issue...."

The F-15B ACTIVE (NASA 836 on the right) and the F-15A chase plane (NASA 837) are shown preparing for in-flight during a July 1996 research flight. The U.S. Airforce developed these F-15 Eagles for air superiority fighters in the late 1960s, and the Eagle first flew in July of 1972. They are expected to be in service with the U.S. Air Force until 2025.

The United States "basically says it has the right to restrict the use of space to only its allies," Clark said. Adds Jeffrey G. Lewis, an arms control expert at the public policy organization, New America Foundation, "Much of the world was appalled at the tone of the policy. One British newspaper columnist basically said it made space the fifty-first state."

In that context, some experts say, the Chinese test was an effort to force the issue, to show the United States the potential consequences of refusing to negotiate a favorable treaty on the military use of space. "The U.S. was restricting all these arms treaties," said Michael O'Hanlon, a senior fellow in security studies at the Brookings Institution, a Washington, DC, think tank. "For the

Chinese, [the test] was an effort to deal from a position of strength."

GlobalSecurity's John Pike believes China may have another rationale for flexing its space muscle: Taiwan. China has long yearned to reabsorb the breakaway island state, which the United States has pledged to defend. In the short term, Pike said, China has only two strategies that could lead to a takeover of Taiwan. It could bluff the United States in a nuclear confrontation, or it could try something altogether different: fire medium-range missiles from mobile launchers, just as it did in the 2007 test, and take out America's low-flying imaging satellites. Doing so might blind U.S. military planners long enough for Chinese military forces to gain a foothold on the island.

"The Chinese stage these big amphibious exercises off Taiwan all the time. One day, maybe it'll be real," Pike said. "Either the United States will get there quickly enough to stop them, or the Chinese will win the race and there won't be enough American political resolve to kick them out. All the Chinese would need is time." A half dozen sats, Pike said, is all that it would take. "Those satellites are low-hanging fruit. It's a no-brainer."

In that scenario, the ASAT test was not really about China showing the United States its capability. It was about China confirming that its own war plan is feasible.

The Aftermath

The long-term ramifications of the test will take years to play out, but for now, few observers think China scored any gains. "It was a mistake," the Brookings Institution's O'Hanlon says. It fueled American hard-liners who want to restrict American technological cooperation with China.

And "it doesn't help make China's case that it isn't a threatening military power," adds GlobalSecurity's Vick. "It is a threat, and the test showed that."

Whether the United States suddenly accelerates its ASAT capability beyond the testing phase remains to be seen. The country is in the midst of wars in Afghanistan and Iraq; budgets are already tight. Russia is not perceived as a threat, and China has only sixty satellites, few of which are worth shooting down.

In fact, America's most robust ASAT weapons were not designed for destroying satellites at all. They are missiles developed and operated by the Missile Defense Agency, formerly known as the Strategic Defense Initiative. All U.S. ballistic missiles are actually dual-use, and while their ability to shoot down incoming rockets has been proven only in tests, it would be easy to direct them against any low earth orbit satellite. Twenty-four Missile Defense Agency missiles are operational in Alaska and California—far more than would be needed, Pike said, to handle any immediate ASAT needs. There is, he said, "just nothing to shoot at."

For now, that is. The militarization of space has long been debated. With one blown-up old weather satellite, China has made the prospect of a new arms race far more likely. It showed the world that it is willing to go toe-to-toe in the final frontier.

U.S. SAT KILL ARSENAL

For the past forty-five years, the U.S. military has been developing weapons designed to destroy enemy satellites. Here are four of the most notable:

1962

Perhaps the most spectacular test, known as Starfish Prime, occurred on July 9, 1962, when a 1.4-megaton nuclear warhead was detonated 250 miles above the Pacific Ocean. Radiation and electromagnetic pulses disabled at least six satellites and created an eerie, artificial glow for twenty minutes.

1985

2. The United States conducted its only direct antisat test in 1985, when an F-15 climbed to 80,000 feet, then fired a three-stage missile. It caught up to a 17,500-mph Solwind solar observation satellite at an altitude of nearly 300 miles over Vandenberg Air Force Base in California.

1997

3. In 1997 the U.S. Army hover-tested a kinetic energy weapon that could act like a space-age fly swatter. The kill vehicle extends a Mylar sheet to disable satellites without destroying them. The military found the technology too messy and eventually canceled the program.

2005

4. The XSS-11 microsatellite isn't an antisat—but it could be. First launched in April 2005, the spacecraft is designed to circumnavigate targets and relay diagnostic data. Some experts note that it wouldn't be difficult to reprogram the craft to ram into enemy targets.

SURVIVAL TIPS: ZERO GRAVITY

Zero gravity takes a toll on the human body—especially its bones and muscles—and astronauts need time to recover. To survive long amounts of time in zero gravity, astronauts must pay critical attention to their health before, during, and after long visits to space.

BEFORE: Astronauts spend months training for space travel. Fitness is important, but they must also prepare by being placed in motion devices designed to provoke and simulate space sickness. It's the least favorite part of most astronauts' training, as they are placed in a spinning chair and encouraged to ride as long as possible.

They must train in zero-gravity planes, where they practice eating, drinking, and going to the bathroom to acclimate them to space conditions.

DURING: Muscle mass decreases in space by about 5 percent per week, and bones atrophy at a rate of about 1 percent per month. For long space trips, without exercise, the effects of zero gravity can be disastrous to the human body.

Astronauts work out on cycle ergometers and treadmills and with rubber cords, and perform resistance exercises such as squats and heel raises to help offset the loss of bone and muscle.

AFTER: Upon returning to earth, astronauts are usually thirsty because their bodies sense that they don't have enough blood in their blood vessels. And it usually takes astronauts months—sometimes even years—to recover their bone mass. Exercise is the key to returning to normal. They also need to regain their balance in earth's gravity, so treadmills with moving bases and projected screen images are used to simulate space conditions.

SHIPWRECKS

Nearly every year the Bureau of Labor Statistics ranks commercial fishing as America's most lethal job. Adjusted to the size of the workforce, the 2008 fatality rate for U.S. fishermen was five times that of truck drivers, eight times that of police officers and 19 times that of firefighters.

It's no coincidence that the number of lost boats and lives is far higher for fishing than for any other type of waterborne industry. Passenger ferries, cargo ships and virtually all other commercial boats are held to much higher regulatory standards. However, federal standards and laws can't predict or protect against pilot error, mechanical failure, or natural disasters. Here are four shipwrecks that might have been prevented.

The RMS *Titanic*, the largest passenger steamship in the world set off for New York City from England on April 10, 1912. But just four days into the crossing, near midnight on April 14, *Titanic* struck an iceberg, causing it to sink just hours later in the icy waters of the Atlantic Ocean. Of the 2,227 people on board, 1,517 perished, making it one of the deadliest maritime disasters in peacetime history.

Old salts still speak of it in clichés: "Go to Halifax," or "I'll blow you to Halifax." To youngsters, it's a legend. For both, the reference is to the biggest man-made explosion prior to the development of the atomic bomb. On the morning of December 6, 1917, two ships, one loaded with more than 2,500 tons of explosives bound for the war in Europe, collided in Halifax Harbor. The resulting explosion killed nearly two thousand, injured another nine thousand, and blinded two hundred.

Halifax Harbor was peaceful on the fateful morning. It could have remained so if the *Mont Blanc*, inbound to anchorage, had been allowed to follow normal right-hand traffic, and the *Imo*, outbound to the Atlantic, had done the same.

10 MOST DEVASTATING SHIPWRECKS

Location	Date	Deaths
1. Dona Paz	1987	4,375
2. Halifax Explosion	1917	1,950
3. Joola	2002	1,863
4. Sultana	1865	1,800
5. Titanic	1912	1,517
6. Empress of Ireland	1914	1,012
7. Estonia	1994	852
8. Eastland	1915	845
9. Birkenhead	1852	460
10. Mary Rose	1545	400

Instead, stubborn insistence on right-of-way put both ships on a collision course near the coast of the town of Dartmouth and set the scene for the terrific explosion.

The SS *Mont Blanc* had arrived with its perilous cargo from New York at dusk the night before, anchoring for the night in Halifax Harbor, south of the narrows. In the morning it would move into Halifax to pick up coal.

The SS *Imo*, a Norwegian liner, was also docked in Halifax to pick up coal, en route to New York to collect relief supplies for war-stricken Belgium. Its captain was angry: the ship had been forced to spend the night because coal delivery had been delayed. So early the next morning, when the *Imo* finally started south into the narrows, it accelerated to a brisk 7 knots. Meanwhile, the *Mont Blanc* had also started up, hugging the right shore and proceeding northward cautiously because of its deadly cargo. The ship was not flying a red flag indicating it carried explosives—such a flag was only required when loading. Tragedy was minutes away.

8:35 AM: On the northbound ship, the men on the bridge were appalled to see the southbound *Imo* fail to take the normal course… out to sea. "That fool is aiming to come down in our water," the harbor pilot said to Captain Aimé Le Medec. "Better give him a whistle." A signal was blown. This demanded some reply—an acknowledgement of port-to-port passing—but to the disbelief of the men on the *Mont Blanc*, *Imo* gave two blasts, indicating a starboard passing.

8:42 AM: When the *Mont Blanc*'s pilot realized the *Imo* was going to try to squeeze between it and the shore, he shouted, "Stop our engines!" Then, in a desperate maneuver, he swung the *Mont Blanc* hard to the left. But it was too late. The *Imo*, instead of continuing on course, blew three blasts,

The city of Halifax, Nova Scotia, Canada, before the SS *Mont-Blanc,* fully loaded with wartime explosives, collided with the SS *Imo*.

The ensuing blast obliterated all buildings and structures within a 500-acre radius and caused a tsunami in the Halifax Harbour.

CITY OF HALIFAX

The blast reduced the *Mont Blanc* to splinters and was so forceful that the ship's cannon was launched to land nearly four miles away. A large portion of the ship's anchor, weighing more than 1,000 pounds, was discovered two miles from the blast, in the opposite direction as the cannon. In the immediate vicinity of the blast, the north side of the city of Halifax and part of Dartmouth were demolished and disintegrated by fire, including homes, businesses, schools and churches. Windows were broken in a 50-mile radius, and shock-waves from the blast were felt as far as 300 miles away.

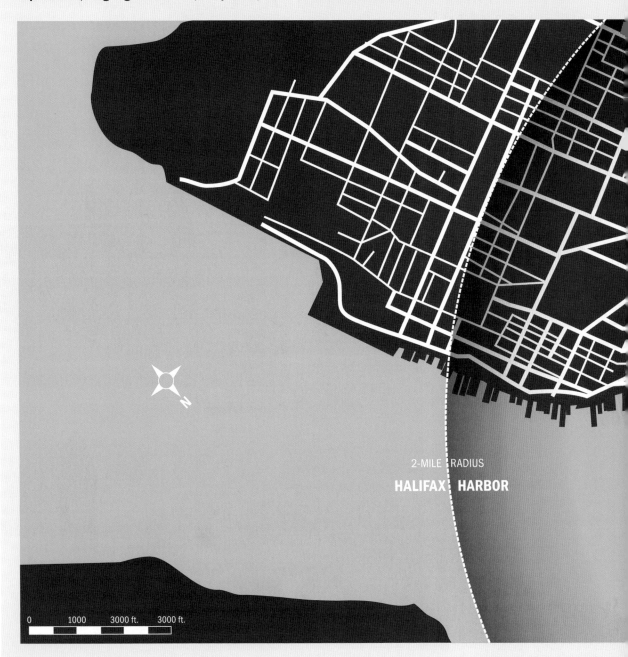

2-MILE RADIUS

HALIFAX HARBOR

0   1000   3000 ft.   3000 ft.

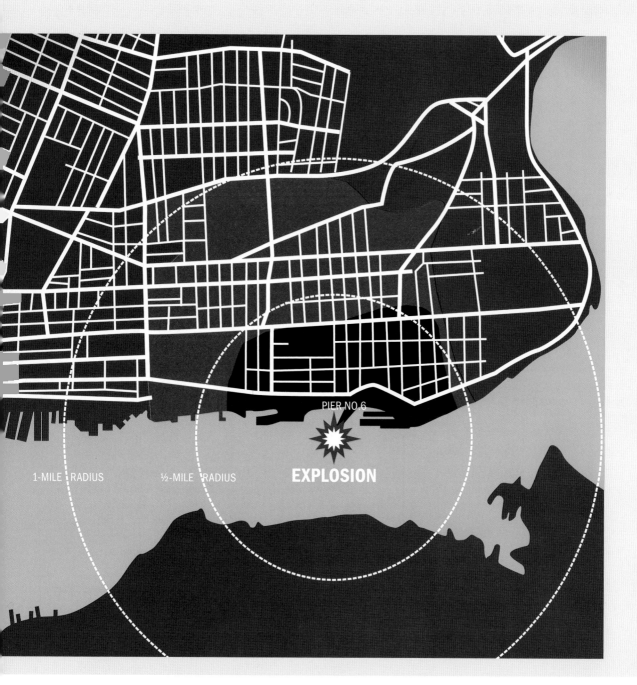

PIER NO.6

EXPLOSION

1-MILE RADIUS

½-MILE RADIUS

"Tragedy was minutes away."

signaling that it had reversed its engines. But the reversed engines threw its own bow to its right. The *Imo* continued drifting close; collision was imminent.

8:45 AM: The bow of the *Imo* sliced 10 feet into the bow of the *Mont Blanc.* Barrels of gasoline smashed open and flooded the deck. Some fell overboard. Then the *Imo*, with its engines still in reverse, backed away, and sparks from the rent metal ignited the spilled fuel. The *Mont Blanc*, now on fire, continued drifting to port and came to rest near the docks in downtown Halifax.

8:55 AM: As thousands lined the shore to watch, unaware of the impending catastrophe, cowards and martyrs played out a grisly scene. The crew of

the *Mont Blanc*, who knew of the ship's deadly cargo, fled in lifeboats to Dartmouth shore and hid in the nearby woods. Meanwhile, tugboat crews and others from nearby vessels, who saw only a ship in distress, tried valiantly to tow it away from the docks and prevent the flames from spreading further. Some intrepid seamen even climbed aboard the stricken *Mont Blanc* to help quench the fire. Worse, factory windows filled with spectators. Tramcars slowed down so passengers could get a better look. Most would perish in the disaster: the drifting ship was the greatest time bomb mankind had ever assembled.

9:06 AM: Twenty-one minutes after the collision, the *Mont Blanc* blew up, devastating Halifax, its citizenry, and the surrounding territory. The explosion blew the bottom right out of the harbor; huge rocks landed on shore. Part of the *Mont Blanc*'s heavy iron anchor, weighing half a ton, came down 2 miles away. An officer on one nearby ship was blown from its deck and landed, bruised and naked, on Ford Needham Hill, a half mile away. Water blown out of the harbor hit a ship at sea so hard its crew thought it had been struck by a mine. Whole blocks of Halifax and Dartmouth were knocked flat. Much of what was left was on fire. Many were killed instantly—others died of their injuries. And it was still not over. The explosion, which happened almost in the middle of the narrows, blew so much water out of the slim channel and out to sea that when it rushed back in it came in waves 30 feet high, a man-made tidal wave that pushed two blocks deep into both towns.

Much of what was left was on fire. Many were killed instantly—others died of their injuries.

Nurses prepare bandages for emergency relief on December 6, 1917. Nearly 2,000 people were killed in the blast, and 9,000 more were injured, most of them seriously. The majority of the wounds were permanently debilitating, as hundreds of people were blinded by flying glass from shattered windows. The destruction was so great that later on, Captain Le Medec and his pilot were brought up on charges. But after a lengthy time, the charges were eventually dropped and blame for the Halifax explosion has, to this day, never been officially assigned.

The Aftermath

When the survivors dared lift their heads, Halifax was in shambles, a horrible sight to behold. The explosion had laid waste to the city. Thousands of homes, factories, and office buildings were destroyed without a trace. Trains were blown off their tracks. There were fires everywhere. A few ships, including the *Imo*, were badly damaged but still floated, their crews dead or dazed. Both the *Imo*'s captain and the harbor pilot aboard were killed instantly. Later, charges were brought against Captain Le Medec and his pilot, but after a lengthy period of legal maneuvering, the charges were dropped. To this day, blame for tragedy has never been officially assigned.

ANDREA DORIA

**Offshore the New England Coast:
July 25, 1956**

Like the collision of the *Imo* and the *Mont Blanc*, mystery surrounds one of the more famous shipwrecks of the twentieth century. In the more than fifty years since its sinking, the *Andrea Doria*, like the *Titanic* and the *Lusitania*, has become a legend. Named for a sixteenth-century Italian admiral, the magnificent 29,000-ton vessel was built in 1951 at a cost of $27 million. Its walls were hung with fine paintings and silver plaques, and its salons were richly appointed.

On its last trip, the *Andrea Doria* carried 1,709 passengers and crew. Its cargo manifest listed, in addition to nearly 1,800 bags of mail, such items as cases of rare wines, crates of antiques, and a fleet of fine cars.

At 12,165 tons and only 254 feet in length, the *Stockholm* was a more modest ship. Built eight years before for the Scandinavia-United States run, it had a heavily reinforced bow for ice-breaking in northern harbors.

On the night of July 25, 1956, the two liners were moving through a blanket of fog off the New England coast, the *Stockholm* eastbound and the *Andrea Doria* westbound, less than a day from making port in New York. This was the era of the ocean liner—before the jet took over international travel—and in summer the shipping lanes across the Atlantic were teeming with ships.

At 11:22 PM, the Swedish and Italian liners collided at a point about 50 miles off Nantucket Island. According to passengers aboard both ships, the other vessel appeared suddenly as clusters of lights emerged from the fog.

A row of first-class cabins on the *Andrea Doria* was demolished by the *Stockholm*'s prow. A man from Brooklyn, New York, staggered out of a bathroom to find that the cabin furnishings and his wife, who had been reading in bed, had simply disappeared through a gaping hole. Others were

luckier. A fourteen-year-old girl had been scooped out of her bed by the *Stockholm*'s prow and was found injured but alive on the deck of the Swedish ship.

The final casualty count showed 45 persons dead and missing from the *Andrea Doria* and 5 dead or missing from the *Stockholm*. After Captain Harry Gunnar Nordenson had determined that the *Stockholm* was seaworthy, he dispatched lifeboats, picking up 533 survivors before heading back to New York.

SOS messages brought a number of ships to the rescue. A U.S. military transport and a destroyer escort, at sea for gunnery practice, picked up nearly 250 survivors. But the ship that really capped the rescue effort was the *Ile de France*. Headed east when the SOS arrived, it swung around and arrived at about 1:30 AM, in time to rescue 753 persons.

At 9:30 AM on July 26, Captain Piero Calamai finally left the stricken ship. A Coast Guard cutter soon radioed a final report: "SS *Andrea Doria* sank at 10:09 AM."

The Aftermath

Responsibility for the collision remains somewhat controversial. Both ships, it was claimed, had used radar to track each other. Why this precaution failed was never satisfactorily explained.

Captain Nordenson, who was found blameless by his company, claimed that the *Andrea Doria*'s crew could not handle radar gear properly. Captain Calamai charged that the *Stockholm* was traveling east in a westbound shipping lane and should have been farther south. Captain Nordenson maintained that the Italian ship had made a last-minute panic turn right into the path of his ship. The Italian

The Italian passenger liner SS *Andrea Doria* sinks after colliding with the Swedish ship Stockholm in heavy fog off Nantucket Island. The Italian Line ocean liner bound for New York City collided with the MS *Stockholm* on July 25, 1956, killing 46 people. 1,660 passengers and crew were rescued, however, and the ship sank the following day. Two U.S. ships in the area at the time responded to distress calls and managed to rescue 250 survivors. Soon after, the Ile de France swept back to the site of the collision and managed to rescue another 753 people.

skipper denied this but could offer no proof; the course-recorder equipment and graphs had gone down with his ship. Legal aspects of the controversy finally were resolved when both steamship lines agreed to end claims against each other for ship damage.

A man staggered out of a bathroom to find the cabin furnishings and his wife had simply disappeared through a gaping hole.

SELENDANG AYU

**Breaking on the Alaskan Coastline:
December 6, 2004**

Sometimes a shipwreck has further-reaching consequences, such as the destruction of Halifax. Though the disaster cannot compare in terms of human lives lost, in 1989, the infamous tanker Exxon *Valdez* spilled 11 million gallons of oil into Prince William Sound off the coast of Alaska, killing thousands of endangered animals, hundreds of thousands of seabirds, and millions of fish. It was one of history's worst environmental disasters caused by humans. Fifteen years later, on December 6, 2004, a similar disaster unfolded.

It was early afternoon and the *Selendang Ayu* was 100 miles northwest of Unalaska Island, about 880 miles from Anchorage in the eastern Aleutians, when alarms rang through the ship's bridge. Inside the six-cylinder two-stroke MAN B&W 6S60MC engine, a cylinder liner had ruptured. The captain was faced with a difficult decision: run the engine long enough to find safe harbor and risk destroying the engine at sea, or shut it down, leaving the ship at the mercy of the waves while the crew attempted repairs. The captain chose to shut it down.

Without propulsion, the *Selendang Ayu* coasted for 2 miles until the sea turned it broadside to the waves. The crew tried to run the engine with only five cylinders, but it wouldn't start. As the rising wind began to whistle through the pilothouse railings, the engine room reported to the captain: the *Selendang Ayu* was adrift.

The crew's fate depended on repairing the 11,542 horsepower engine, a two-story-high behemoth that could power a town of thirty thousand people. The men would have to use a movable crane to remove the ruptured 3-ton liner, swap in a new liner, and restart the engine.

The men worked under impossible conditions, the hull rolling wildly as each wave hit. Worse, the northwest wind was blowing the ship toward the rocky

coast of Unalaska Island. At 1 AM on December 7, the captain finally radioed for help.

Sixty-five miles to the northeast, the 283-foot Coast Guard cutter *Alex Haley* took the call from its command center. By the time the cutter reached the scene at 11:30 AM, the *Selendang Ayu* had drifted 25 miles closer to shore. Attempting to pass a towline to the freighter, the *Alex Haley* fired a messenger line from a handheld gun, but the line broke in the 45-mph gusts.

The storm's swells piled higher as the *Selendang Ayu* drifted toward the steep, white-flanked mountains of the coast. Two other tugs arrived from Dutch Harbor, 40 miles away on the other side of the island, but with waves breaking higher than the boats' wheelhouses, neither could get close enough to assist.

The crew had to buy enough time to get the engine started. Once the water grew shallow enough, they lowered an 11.5-ton steel anchor into the churning ocean. The ship settled nose-first into the wind, and for the first time crew members were able to work on the engine without the hull rolling violently. But the anchor didn't hold. In less than two hours it dragged 3 miles along the bottom until it broke.

The struggle to repair the engine neared its final act. By 2 PM, with the winter sun sinking low and the surf breaking on the beach less than a mile away, the crew managed to deploy its second and last anchor. Again the bow swung into the wind. But how long would the anchor hold? If the crew was going to jump ship, the time to try was now.

All afternoon, two Coast Guard HH-60 Jayhawk helicopters based out of Kodiak, Alaska, had circled nearby. Flying conditions were terrible, with winds gusting up to 70 mph and periodic snow squalls cutting visibility to zero. At the best of times, hovering a helicopter is a demanding skill, a

An over-flight photo taken on December 19, 2004, shows the bow and stern sections of the 378-foot freighter MV *Selendang Ayu* near Skan Bay, Alaska. Carrying soybeans from Seattle, Washington to China, the ship ran aground near Unalaska Island, resulting in a massive oil spill. It's estimated that approximately 320,000 gallons of bunker oil were spilled in the wreck. High winds, snow, 40-foot waves and limited visibility made rescue efforts nearly impossible for the crew of the Coast Guard cutter Alex Haley and the two Jayhawk helicopters that arrived on the scene.

never-ending argument between pilot and physics. Hovering in a storm is nearly impossible. But for chopper crews in Alaska, it is practically routine.

The first Jayhawk, piloted by Lieutenant Doug Watson, swooped in and lowered a hoist basket, pulling up nine crew members one by one. But getting the rescue victims onto the slick, pitching deck of the nearby *Alex Haley* proved extremely hazardous. "Once they got the first nine crew members on board, the flight officers on the deck said, 'This is way too dangerous. Don't bring any more,'" remembers Coast Guard spokeswoman Sara Francis, who had been monitoring communications.

"This is way too dangerous."

Efforts to evacuate the 18 crewmembers fell to the U.S. Coast Guard cutter *Alex Haley*, along with two Jayhawk helicopters, which hovered over *Selendang Ayu* in an attempt hoist crewmembers onto the aircraft. A rogue wave broke over the bow of the ship and one of the copter's engine's flamed out before crashing into the sea. Six men lost their lives in the wreck of the *Selendang Ayu*. But it would have been many more were it not for the heroics of the Coast Guard crewmen who put their lives in great peril in order to save the traumatized crew of the ill-fated ship.

Instead, the second Jayhawk ferried nine more crew members from the *Selendang Ayu* to a landing spot on shore. The Coast Guard offered to remove the remaining men, but the captain and seven crew members opted to make a final bid to restart the engine. They kept working, even after the last anchor gave way at 4:30 PM.

Forty minutes later, the freighter shuddered as it hit a submerged rock, which tore a gash in cargo hold No. 4. The mangled steel moaned as seawater roared into the mortally wounded vessel. The ship settled onto the shallow bottom, its deck resting just above the surface. Breakers crashed; the wind howled. Though there were three survival suits aboard, the men wore thin civilian clothing and life jackets. They would last mere minutes in the 32 degree water. The call came through to the circling Jayhawk: go get the crew.

A smaller HH-65 Dolphin helicopter was launched from the *Alex Haley* to provide backup for Watson, who fought gale-force winds as he brought his Jayhawk in low. Copilot Lieutenant Dave Neel kept an eye out for incoming waves while Petty Officer Second Class Brian Lickfield, a flight mechanic, lowered the rescue basket to the wave-swept deck 30 feet below. He motioned for the crewmen to climb in, but they simply stared. "I think they were just terrified," Lickfield later told the *Anchorage Daily News*. "Nobody would move. We were wasting a lot of time. In the end, time became lives."

A rescue diver was going to have to go down. Petty Officer Third Class Aaron Bean descended in the basket, then walked each of the men over to it as the swells grew bigger, some cresting as high as 40 feet. Six of the crew were inside the Jayhawk and a seventh was on his way up when Neel saw a rogue wave headed toward them. He warned Watson, but the pilot had to hold steady as the

basket was pulled inside. An instant later the wave hit. Icy water surged into the cockpit and dragged the struggling chopper from the sky, its rotors splintering as they slammed into the ocean's surface.

The inside of the helicopter was a pandemonium of rushing water, alarm horns, and flashing lights as the Jayhawk rolled over and began to sink. The three Coast Guard crewmen, wearing buoyant survival suits, floated to the surface, where the Dolphin picked them up. Six men from the *Selendang Ayu* weren't as lucky. The seventh was pulled from the churning sea close to death, suffering from hypothermia and injuries sustained in the crash. By the time he reached the Dutch

↑ The shipwreck of the *Selendang Ayu* resulted in the worst Alaskan oil spill since the infamous Exxon *Valdez* in 1989. The shipwreck was a blow to Alaskans, who remembered all too well the environmental damage an oil spill could inflict. High winds and cruel weather made the cleanup effort that much more difficult, and salvage and cleanup continued for years after the wreck.

"The ship hemorrhaged over 320,000 gallons of... heavy fuel oil."

Harbor Clinic, his body temperature had fallen to a dangerously low 78 degrees, but he did survive.

The drama at sea was far from over for the ship's captain and rescue diver Aaron Bean, still huddled aboard the bow of the *Selendang Ayu*, where frigid waves swept the deck. A long hour passed, and then the relentless violence of the sea tore the hull in half, sweeping the aft section hundreds of feet away. The lights aboard the bow died, plunging the two men into darkness. Battered by waves and pummeled by wind, the men held on, believing all aboard the Jayhawk were dead. Though they didn't know it, the Dolphin was fighting its way back from Dutch Harbor. But this time, there was no backup. If the

Dolphin went down, the chance of anyone surviving would be slim.

Thirty minutes passed, then an hour. At last Bean heard the thwack-thwack-thwack of the Dolphin's rotors above the roar of the wind. Hovering in the turbulence, the chopper lowered its basket. Within minutes Bean and the ship's captain were inside the helicopter, winging to safety.

The Aftermath

The wreck of the *Selendang Ayu* not only cost the lives of six men, it resulted in the worst Alaskan oil spill since the Exxon *Valdez*. The ship hemorrhaged over 320,000 gallons of bunker C—heavy fuel oil that can persist for years in cold water and travel hundreds of miles—as well as thousands of gallons of diesel oil that can spread rapidly into thin slicks. High winds and seas hampered cleanup efforts; miles of coastline were befouled, and more than 1,600 seabirds were killed.

Unalaska residents witnessed the ecological repercussions and understood the human loss all too well. "During the fishing season, we all worry about our loved ones," said Unalaska mayor Shirley Marquardt. "When the north wind is blowing, there's nowhere to hide."

While it is true that the effects of some shipwrecks, such as the *Valdez* and the *Selendang Ayu*, can be measured in terms of money lost, oil spilled, or square miles fouled, some shipwrecks can only be measured in the simple and terrible cost to human life.

The carcasses of more than 1,600 birds were recovered from beaches along the shore of Unalaska Island after the *Selendang Ayu* spill, as well as 6 sea otters. Environmental officials estimate that because the ship's heaters were turned off, the fuel thickened in the cold waters, making cleanup efforts more cumbersome. Pictured is one of the birds from the Exxon *Valdez* spill.

When *Selendang Ayu* wrecked, it was an eerie reminder of another oil spill in Alaska—the infamous Exxon *Valdez* in 1989, when images of cleanup crews in bright yellow and orange suits were a nightly fixture on the evening news, bringing the environmental catastrophe into living rooms in living color. Unlike the Exxon *Valdez*, however, the wreck of the *Selendang Ayu* led to the deaths of six men as the U.S. Coast Guard made a valiant attempt to rescue crew members from the sinking ship.

On a warm afternoon in October 2005 at Lake George, New York, a group of forty-seven retirees from Michigan and Ohio climbed—some with the assistance of walkers—aboard the 38-foot tour boat *Ethan Allen*. It was one of the last stops on a week-long fall foliage trip, and many passengers opted for port-side benches, where the sun streamed in beneath a wooden roof through open windows. Jeane Siler, 77, noticed the boat was listing and joked to a companion, "Hey, we might not get away from the dock."

The *Ethan Allen* nonetheless pushed off and cruised through placid water, past green bluffs and historic mansions, to a cove near Cramer Point, where the captain noticed an approaching wake. It struck the hull before he could fully turn the boat, sending passengers sliding down the benches. The boat heeled to port, then capsized completely. "I was swimming for my life," Carole Mahalak, 67, later told the National Transportation Safety Board (NTSB). She struggled through an open window toward the light.

Nearby boaters quickly tossed life jackets into the water, but it was fifteen minutes before two divers, who had seen the boat flip as they taught a scuba class, could attract someone to shuttle them from shore. When they arrived at the site, the divers found twelve bodies in the upended hull. The boat abruptly sank 59 feet to the bottom of the lake as they extracted the last one. Twenty people, nearly half the passengers, had drowned. It was the country's deadliest pleasure boat accident in decades.

It was also one of the most inexplicable. The boat had capsized while carrying a legal passenger load and after operating without incident for more than forty years. The NTSB launched a ten-month investigation that examined everything from the bilge pump to the skills of the seventy-four-year-old captain.

The Aftermath

Investigators learned the *Ethan Allen*, delivered in 1964 as an open-air vessel, had never been retested for stability after receiving its 2,000-pound roof in 1989. The added weight above the deck caused the boat to sit lower in the water and raised its center of gravity by about 7 inches—changes that reduced its maximum safe passenger load from forty-eight to just fourteen. "Even just a small amount of weight up high can really make a difference," says George Borlase, a naval architect who has worked on many ship casualty investigations,

Still, the *Ethan Allen* might have righted itself if not for other factors. An uneven seating arrangement gave the boat a 2.2-degree list. When the wake and the force of the captain's turn increased the tilt enough to cause passengers to slide, they would have had to move less than 1 foot outboard for the boat to capsize. The same shift could have toppled the boat even at harbor: New York's benchmark for calculating maximum loads, set in 1960, was a 140-pound person; the passengers that day, following a national trend, were on average 38 pounds heavier. Jack Spencer, director of the NTSB Office of Marine Safety, says, "It was clear that the boat had several thousand more pounds on it than it should have."

The state responded to the accident by raising its passenger weight benchmark to 174 pounds. The NTSB recommended that the Coast Guard provide guidance to New York and other states that haven't adopted its more stringent small-vessel inspection standards, which otherwise apply only in areas with connections to interstate waters. Even where laws don't require it, Spencer advises boaters who have altered their vessels to hire a professional to check the stability. "That was really the fatal flaw of the *Ethan Allen*," he notes.

ETHAN ALLEN: A DEADLY SEQUENCE OF EVENTS

Angle of capsize: If a boat rolls to one side beyond a certain point of stability, it can capsize. This point is considered its maximum heeling angle.

1964: The *Ethan Allen* initially operated as an open-air vessel. It was approved to carry forty-eight passengers and two crew weighing an average of 140 pounds each. With a high freeboard (plenty of hull above the water) and a low vertical center of gravity, the boat could heel up to 56 degrees and right itself without capsizing.

2005: A 2,000-pound roof and passengers weighing on average 178 pounds raised the boat's center of gravity, and caused it to sit lower in the water. This weakened its righting moment—the springlike force that rights a boat. Anything beyond a 39.8-degree heel was now enough to cause the boat to capsize.

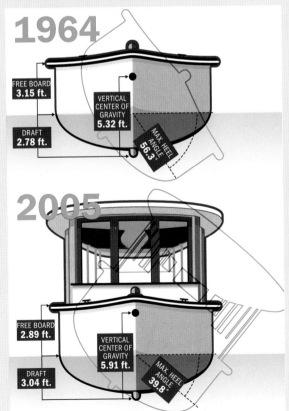

1964

FREE BOARD
3.15 ft.

DRAFT
2.78 ft.

VERTICAL CENTER OF GRAVITY
5.32 ft.

MAX. HEEL ANGLE
56.3°

2005

FREE BOARD
2.89 ft.

DRAFT
3.04 ft.

VERTICAL CENTER OF GRAVITY
5.91 ft.

MAX. HEEL ANGLE
39.8

Wake approaches: As the wake of another boat rolled toward the *Ethan Allen*, an uneven distribution of the vessel's forty-seven passengers was already causing it to list 2.2 degrees to the port side.

Passengers slide: The captain turned the boat to ride out the wake, further heeling it to port. When he failed to hit the wake head-on, the boat rocked even more and sent passengers sliding down the benches.

Boat capsizes: Once passengers shifted a mere 10 inches from their original seating positions, the *Ethan Allen's* maximum heeling angle dropped from 39.8 degrees to only 16.8 degrees. The boat quickly capsizes.

SURVIVAL TIPS: SHIPWRECK PREPARATION

When the water rushes in and the ship is going down, the right knowledge and equipment can save lives. Every boat should have an emergency position-indicating radio beacon.

➤ "When your boat tips over, the automatic beacon dislodges itself, floats to the surface, and sends out a signal," Coast Guard search-and-rescue coordinator Denny Ernster says, "and there's no range limitation on it. Then we know exactly where you are."

➤ If you're waiting for rescuers in cold water, stay still. With hypothermia, blood vessels constrict, reducing the supply of warm blood to the skin. That keeps internal organs warm—which is what you want. But forcing those vessels open by exercising in the water pushes the warm blood to the surface, where it quickly gets chilled.

➤ Finally, keep in mind that most humans are naturally buoyant; we float, but just below the surface. So rest by floating facedown in the water with arms out, scarecrow style. Every fifteen seconds, raise your arms to the surface, then push down. The motion causes your head to rise above the surface long enough for you to take a breath.

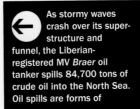 As stormy waves crash over its super-structure and funnel, the Liberian-registered MV *Braer* oil tanker spills 84,700 tons of crude oil into the North Sea. Oil spills are forms of pollution that occur in the environment due to human activity. When liquid petroleum hydrocarbon or refined petroleum products such as gasoline is released, spills can take months if not years to clean up.

INDUSTRIAL ACCIDENTS

The modern age is one of unprecedented convenience and luxury. Electricity courses through houses and cities, providing heat and light. Oil and gasoline power vehicles that allow travel to the ends of the earth, making the world a smaller place than it was just half a century ago. And for many Westerners, all the food they could ever need is as close as the nearest supermarket.

But all this luxury comes at a price: the technology needed to provide many modern amenities is so advanced that it resembles the stuff of science fiction from mere decades ago. On March 24, 1911, the death of 146 garment workers, almost all of them women, at the Triangle Shirtwaist Factory made it the largest industrial disaster in New York City. The women were unable to escape the flames because managers locked the doors on a daily basis to make sure none of the garment workers could sneak out to the stairwells for cigarette breaks. The tragedy led to implication of many of the workplace regulations that are common in businesses today.

On August 17, 2009, the Sayano-Shushenskaya Hydro Station in Russia suffered a catastrophic accident that flooded the turbine rooms and caused a massive transformer explosion. 74 people were killed in the incident, which caused an oil spill of more than 40 tons of transformer oil that spread more than 50 miles downstream of the Yenisei River. It's estimated that the cost to rebuild just the engine room would exceed $1 billion. This accident has forced the world to consider whether hydropower plants are safe.

IMPERIAL SUGAR REFINERY

Explosion in Port Wentworth, Georgia: February 7, 2008

Greg Long has seen his share of disasters. The chief of the Port Wentworth (Georgia) Fire Department is a U.S. Army veteran, former police officer, certified arson investigator, and emergency medic. But nothing in his professional background prepared him for the inferno he faced at the Imperial Sugar refinery on February 7, 2008. "I've seen plane crashes, auto accidents, pileups, and violent crime," he said, "and this was the worst."

The nearly one-hundred-year-old Port Wentworth plant, which sprawls over 155 acres along the Savannah River, is the country's second largest sugar refinery, producing 2.6 billion pounds in 2007, or 9 percent of the U.S. total. Barges bring raw cane sugar upriver to the facility, where it undergoes a nine-step refining and concentrating process before being packaged as Dixie Crystals.

The refinery was operating normally until just after 7 PM, when a strong but localized explosion rocked the warehouse where the product is packaged. The blast occurred where conveyor belts bringing sugar from 100-foot storage silos enter the 6.3-acre structure.

Vibrations shook loose sugar dust that had accumulated on conveyors, light fixtures, and rafters, filling the air with tiny particles. What workers inside did not realize was that the sugar dust in the air had turned the building into a bomb that only needed an igniter—something as innocuous as static electricity—to detonate.

Dust explosions occur when combustible material with a high surface area, suspended in the air in a confined space, meets an ignition source—anything from overheated ball bearings to an electric spark. Particles of 0.42 millimeters or less (small enough to float in air) are explosive; sugar dust can be as small as 0.03 millimeters. The disruption of a layer of dust as shallow as $1/32$ of an inch can start a deadly chain reaction.

Five minutes after the initial explosion, the airborne dust in Imperial's warehouse ignited. The boom rattled windows 7 miles away. The overpressure of the expanding fireball collapsed the walls and roof. Burning dust particles, propelled by the force of the explosion, blew along the conveyor belts and spread the inferno to two silos holding sugar. "The tops of the silos shot up and flew a good half mile," said Michael Godbold, emergency management coordinator for the nearby Garden City.

"The tops of the silos shot up and flew a good half mile…."

The silos continued to burn, reaching 4,000 degrees—too hot to be extinguished by water dumped from helicopters. Specialists from Texas mixed river water with a fire-suppressing foam and sprayed it onto the top of the silos. On day six, the molten sugar inside finally cooled.

Despite the damage, Imperial was able to get up and running again. The company town of 3,400, however, did not recover as quickly from the disaster's human cost: 13 workers died, dozens were injured, and 15 were admitted to a burn center, some for months.

The Aftermath

Refineries often voluntarily install safety equipment. Imperial's plant had dust collectors to remove particles from the air in places where material was being moved and vent panels designed to blow out under the pressure of a blast to localize structural damage. Ironically, dust collectors are frequently a location for explosions: one atop Imperial's roof had blown up just weeks before the fire. At some

Smoke still billows from a section of the Imperial Sugar Company plant after an explosion ripped apart the plant on the Savannah River on Friday, February 7, 2008, in Port Wentworth, Georgia. Thirteen people were killed in the blast and 42 more were injured after a dust explosion occurred in the center of the refinery where bagging and storage facilities were located. Witnesses reported seeing flames shoot up several stories high. The Chemical Safety Board issued a final report in September 2009, citing inadequate equipment design, maintenance, and housekeeping practices as the causes for the fatal sugar dust explosions.

factories, infrared detectors used to spot flames or sparks in air ducts or conveyor systems are mated with explosion-capture equipment designed to squelch blasts at their inception.

The U.S. Chemical Safety and Hazard Investigation Board (CSHIB), which investigates industrial accidents and makes safety recommendations, analyzed the Imperial Sugar Refinery explosion for a year and a half to determine what triggered the initial blast. In September 2009, the Chemical Safety Board's final report cited inadequate equipment design, maintenance, and housekeeping practices which led to sugar dust explosions.

The complex chemical and industrial processes that occur in refineries make them dangerous, regardless of what they are refining. Processed sugar is a luxury, but much of the Western world has come to consider oil a necessity. With a product as volatile and unstable as oil, the smallest misstep can spell disaster.

TEXAS CITY REFINERY EXPLOSION, TEXAS

Safety Failure on March 23, 2005

For safety inspector Charles Ramirez, things got off to a good start. His team of specialized contractors from Houston-based J E Merit Constructors had just polished off a celebratory lunch held on the grounds of BP's massive Texas City, Texas, refinery. The 1,200-acre facility processes up to 460,000 barrels of raw crude oil a day, and the contractors had just wrapped up their part in its complicated, nine-week "turnaround," or scheduled maintenance cycle, accident-free.

What he didn't realize as he hustled across the yard was that, just 100 feet behind him, the huge steel isomerization unit was being restarted after two weeks offline. The "isom" unit, which boosts the octane level of gasoline, was about to cause the deadliest U.S. refinery disaster in a decade.

The most dangerous time for an oil refinery isn't when it is running, but when it's in transition. During a refinery turnaround, some thirty thousand separate procedures are performed. Dozens of people are required to move volatile contents safely out of and into position when the isom unit is coming back online.

As workers restarted a component of the unit, abnormal pressure built up in the production tower, and three relief valves opened to allow highly volatile gasoline components to escape to the 10-by-20-foot "blowdown" drum. But so much fuel flooded into the drum that its capacity was rapidly exceeded. Liquid and vapor shot straight up the 113-foot vent stack into the open air.

There was no sound...
not even a siren.

Witnesses saw a cloud of vaporizing fuel geyser out of the stack and cascade to the ground. One person reported hearing a desperate call crackle over a handheld radio: "What is this? Stop all hot work! Stop all hot work!"

But too much equipment was running to shut it all down. As vapors were sucked into its engine, an idling pickup at the base of the tower began to rev up. Somewhere in the cloud of fumes, perhaps in the truck's engine, a spark touched off the gas and ignited a firestorm.

When a cloud of highly flammable material is ignited, two events occur almost instantaneously, producing two audible blasts. First, an initial flash consumes all the available oxygen, creating a giant vacuum. Once the suction brings in fresh oxygen, the combustibles explode into a well-fueled inferno that flings a shock wave in front of it.

At close range, this supercompressed wall of air is actually visible as it rockets outward at more than 1,000 feet per second. Ramirez said he saw it just before it blew him to the ground. But his colleagues would have had no warning before it slammed into the flimsy frames of their trailer offices. Eleven of Ramirez's teammates were killed instantly by the blunt force of the shock wave. A fireball then rolled over the shattered trailers and melted nearby portable toilets.

Investigators now suspect there may have been as many as five separate explosions, in rapid succession, including one directly beneath the trailer Ramirez had just left.

Almost a mile from the explosion, BP retiree Shera Shurley was watching TV in her mobile home when its windows blew in. She ran outside to escape. Standing in her driveway, she looked at the swirling black cloud climbing into the sky. There was no sound, she remembers, not even a siren.

Texas City's emergency services crews began rolling moments after the isom unit shattered. BP maintains its own fire brigade and has a mutual response plan with the brigades of the other two Texas City oil refineries, owned by Marathon Ashland Petroleum and Valero. They get plenty of practice: according to Texas City Fire Department chief Gerald Grimm, BP had thirty fire alarms in

HOW A REFINERY WORKS

Refineries separate raw crude oil into its various components, called fractions, by taking advantage of the distinct boiling point of each. The process begins with fractional distillation, when crude oil is heated to about 720 degrees. Hot liquid and vapors enter a distillation column where the vapors cool as they rise, condensing on collection trays at different heights. These liquids, such as naphtha and kerosene, may then be diverted to other units for further processing. Each fuel is made of a distinct chain of hydrocarbons, and manipulating these molecules produces different petroleum products. Cracking units and coker units break large chains into smaller ones to create medium-weight and heavy fuels. Alkylation units combine short chains, forming mainly aviation gasoline. Isomerization units rearrange the structure of molecules to turn naphtha into high-octane gasoline.

2003 and twenty-seven in 2004, although he said this was no more than other plants of a similar size.

Soon seventy-five local, regional, and industrial emergency response units surrounded the site, where walls of water erupted from "monitors"— strategically located water cannons, each capable of hurling up to 1,500 gallons per minute. The thwack of rotors could be heard pounding through the thick smoke overhead: first on scene were news choppers, followed by a Life Flight helicopter from Memorial Hermann Hospital in Houston. Just twenty minutes after the accident, the airspace had become so crowded that the Federal Aviation Administration declared a no-fly zone 3,000 feet high and 3 miles wide.

At the end of the first hour, the fire had been contained, and within two hours, it was nearly out. Only then did the tally sink in: fifteen dead, over one hundred injured. Of the fatalities, more than two-thirds worked for Ramirez's team and had nothing to do with the unit that exploded.

The Aftermath

Texas City knows industrial facilities and their dangers. Often referred to as Toxic City, it is home to four chemical plants and three refineries. The sprawling BP complex, built in 1934, is the third largest of 149 petroleum refineries nationwide. At night it glows like a forested landscape of steel Christmas trees, strung with flickering safety lights. Since records were kept in 1971, there have been at least nine other accidents at the refinery that injured or killed workers, but the explosion on March 23 was by far the most destructive.

In the weeks following the accident, BP's operations came under intense scrutiny. Blowdown drums are a common feature at refineries, as are

Workers sift through debris at the BP facility in Texas City, Texas, 35 miles south of Houston, on March 24, 2005, after a fire and explosion that killed 15 workers and injured more than 170. The explosion occurred in the plant's isomerization unit when hydrocarbon flow to the blowdown drum was overwhelmed, sending liquids over the top of the stack.

towers used to release evaporating gases. Most tower vents, however, include a flare system, a sort of pilot light that ignites potentially hazardous vapors as they funnel out. In 1992, the Occupational Safety and Health Administration (OSHA) mandated that the Texas refinery switch to a flare system. Amoco, which merged with BP in 1998, appealed and OSHA withdrew the request; the refinery continues to use the stacks that allowed gases to escape.

The location of the temporary trailers has also been questioned. BP rules allowed trailers within 350 feet of refining units (although at least two were within 150 feet), provided they receive site-specific analysis. BP has since mandated trailers be located at least 500 feet away. Some other refiners take the additional precaution of requiring nonessential personnel to be evacuated when units like the isom are being brought on line. According to BP spokesman Hugh Depland, BP has no such requirement.

OSHA and the CSHIB both conducted investigations. According to Don Holmstrom, an investigator with the U.S. Chemical Safety and Hazard Investigation Board, his agency looked into whether the fuel had been heated too quickly, which could have led to the pressure spike in the tower, and whether all outflow valves had been working properly. In the end, the CSHIB concluded that BP failed to heed or implement safety procedures that had been recommended before the explosion.

Ramirez, who survived the explosion, was left wondering if an evacuation order that might have saved his colleagues was ever passed along. His boss, Eugene White, might have known, but he died when his trailer office was demolished by the shock wave.

A week after the accident, workers wearing a who's who of petroleum industry caps gathered between shifts in the Texas Tavern. One said a buddy had quit; he'd been having lunch with his wife in a minivan when the plant blew right in front of them. Someone else noted the date: exactly one year since an explosion in another unit at the plant. Soon the bar grew crowded with men drinking longneck Buds and shooting pool. Down the street at the refinery, skeletal cranes, shrouded in fog, continued to pick over the rubble.

HOW THE ACCIDENT HAPPENED

According to the Chemical Safety and Hazard Investigation Board, computerized records from the control system equipment indicate pressure inside the production tower rose rapidly from 20 to 60 pounds per square inch. This triggered three pressure-relief valves to open for six minutes, discharging enough fuel into the blowdown drum to overwhelm the system. Petroleum could not be recycled back through the refinery quickly enough, forcing liquid and vapors up the 120-foot stack. As fuel settled to the ground, it ignited in a blast strong enough to rip the roof off a benzene storage tank 300 yards away. Investigators discovered that a 6-inch drain leading to the plant sewer had been chained open. Fumes traveling under the refinery may have fueled one of what is believed to have been five explosions.

MONTCOAL MINE DISASTER

**Underground Disaster in West Virginia:
April 5, 2010**

A tremendous underground explosion on April 5, 2010, in Montcoal, West Virginia, killed more than two dozen miners in the country's worst coal mining disaster since the Finley Coal Company explosion on December 30, 1970. Though some miners were killed in the blast, officials believe methane gas released into the drilling shafts accounted for more deaths and hindered early rescue efforts.

Don Blankenship, then chief executive of Massey Energy, which owned the mines, said that mine crews, who were investigating a carbon monoxide alarm, discovered that there had been an explosion deep down in the mine. Elevated levels of carbon monoxide and methane gases put initial rescue efforts on hold, and the crews attempted to ventilate the mine by drilling several holes around the shafts.

The Aftermath

Communication can save lives. When first responders arrive at a mine disaster or a building collapse, communicating with victims or other rescuers is usually impossible because radio waves can be blocked by metal, earth, and stone. Even sites with low-frequency emergency radio systems suffer from slow data transfer that restricts use to simple text messages. But a novel system being developed by Ferro Solutions of Woburn, Massachusetts, transmits voice signals with magnetic waves that travel through solid matter more easily than do radio signals.

Typical radios send signals on electromagnetic waves that oscillate at specific frequencies. Engineers at Ferro developed a portable communicator that translates these undulating waves into signals carried by magnetic fields with

The Upper Big Branch Mine disaster on April 5, 2010 at Montcoal, West Virginia was the worst in the United States since 1970. Twenty-nine out of the 31 miners on the site were killed when an explosion from an unknown source occurred. Officials believe that a spark from a mantrip ignited a site that had already high methane levels. The high methane levels delayed the initial rescue attempts.

resonant frequencies that compatible radios can pick up through hundreds of yards of obstructions. The digital voice signal first passes through a transducer made of a composite that melds piezoelectric material (which generates a voltage when it deforms) with a magnetostrictive metal (which changes shape when it's exposed to a magnetic field). The transducer converts the radio signals into magnetic ones; another reverses this process when the signals reach the other radio.

Being able to communicate with trapped miners can shave critical hours of futile searching if a location can be determined through the use of these portable communicators.

People all over the world depend upon electricity, and society goes to great lengths to produce it. Electricity-generating facilities are huge, unbelievably complex places staffed by highly-trained specialists and technicians. Nuclear plants are more closely watched, and the world soon knows when things go wrong—like they did at Chernobyl, Ukraine (then the USSR), in 1986, or at Three Mile Island, Pennsylvania, in 1979. But hydroelectric facilities harness the awesome power of billions of gallons of water and have the potential to be almost as deadly.

Just before 8 AM on August 17, 2009, workers on the morning shift stepped off a clattering Soviet-era tram and made their way past security and into position at the Sayano-Shushenskaya hydroelectric power plant in south-central Siberia. In the 950-foot-long turbine hall, custodians mopped the stone floors and supervisors handed out assignments. On the roof, a technician began installing a new ventilation system. Above him soared a concave dam eighty stories high and more than half a mile wide at its crest. When operating at full capacity, the plant's ten interior penstocks funneled water from the reservoir behind the concrete barrier to the hall below him, where it tore past the blades of ten turbines, spinning them with tremendous force before being flushed out of the hydro plant and down the Yenisei River.

Completed in 1978, the Soviet-era hydro station is Russia's largest, with enough output to power a city of 3.8 million. It was undergoing extensive repairs and upgrades that morning, so more workers were in the hall than usual: fifty-two on the main floor and another sixty-three down in the bowels of the plant. Nine of the ten turbines were operating at full capacity—including the trouble-some Turbine 2, which had been offline but was pressed back into service the previous night when

electricity production dropped because of a fire at the Bratsk power station 500 miles to the northeast. A few minutes into his shift, the technician felt the roof begin to vibrate. The vibrations grew louder and gradually turned into a thunderous roar. Alarmed, he scrambled off the roof.

At 8:13 AM, two massive explosions rocked the hall. Security guard Aleksandr Kataytsev told English-language news station RT that he was one level below the turbine hall when he heard "a loud thump, then another one, like an explosion—and then the room went pitch-black."

Turbine 2—a 1,500-ton piece of machinery topped by a power generator—blasted through the floor and shot 50 feet into the air before crashing back down. The penstock water that had been spinning the turbine geysered out of the now-vacant shaft at a rate of 67,600 gallons per second. Like a massive industrial water jet, it tore down the metal joists over Turbines 1, 2, and 3; the roof there crumpled like aluminum foil and collapsed in a tangle of glass and metal.

The roof there crumpled like aluminum foil and collapsed in a tangle of glass and metal.

Water continued to pour into the hall, flooding its lower levels and eventually submerging other turbines. The plant's automatic safety system should have shut down the turbines and closed the intake gates on the penstocks at the top of the dam, but Turbines 7 and 9 still operated at full speed, in excess of 142 revolutions per minute, triggering the crackling short circuits that darkened the plant. Amateur video footage taken downstream at the time of the accident shows

The official report on the Sayano-Shushenskaya hydro incident states that the accident was caused primarily by vibrations of Turbine No. 2, which led to fatigue damage of its mountings. The majority of bolts examined in the investigation were noted to have had fatigue cracks. The No. 2 Turbine was the plant's power output regulator on the day of the accident. Turbine 2, which weighed more than 1,500 tons, blasted through the floor and shot 50 feet into the air before it came crashing back down. Many employees feared a complete collapse of the dam and quickly began calling friends and relatives who lived downstream, urging them to head for the mountains.

bright flashes and a huge explosion in the vicinity of Turbines 7 and 9 as a wall of water spews from the structural breach near Turbine 2.

As the water level rose, employees stampeded toward the main entrance. Fearing a total collapse of the dam, many phoned relatives downstream and urged them to seek shelter in the surrounding Sayan Mountains. Among the fleeing workers were several supervisors in charge of safety and emergencies, which added to the confusion. On the fourth floor, shell-shocked midlevel operators telephoned up the chain of command for a contingency plan. No one answered.

Using his mobile phone as a flashlight, security guard Kataytsev found his way to an exit and made for higher ground. At the crest of the dam, he and several other employees struggled to manually close the penstock intake gates. By 9:30 AM they had sealed all the gates, and the destruction below ceased.

In the wake of the accident, rescue crews mobilized to search for survivors. RusHydro, the partially state-owned utility company that operates Sayano-Shushenskaya, assembled four hundred employees to pump out the flooded turbine hall and pick through the twisted debris. Russian president Dmitry Medvedev dispatched Sergei Shoigu, his emergencies minister, and Sergei Shmatko, the energy minister, to oversee rescue efforts. Environmental clean-up crews attempted

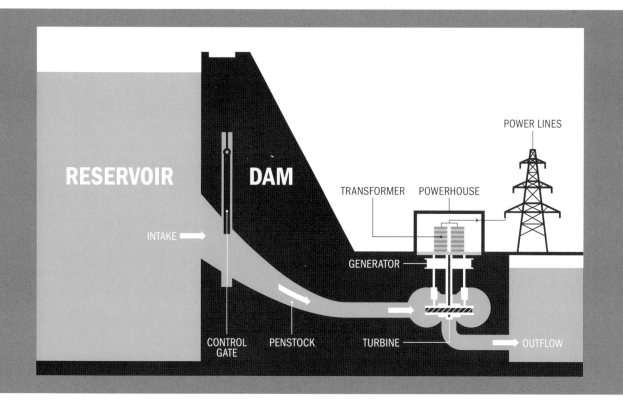

RESERVOIR

DAM

POWER LINES

TRANSFORMER   POWERHOUSE

INTAKE ➡

GENERATOR ─

CONTROL   PENSTOCK   TURBINE ─   ➡ OUTFLOW
GATE

to contain the oil spill that stretched 50 miles down the Yenisei River and killed 400 tons of fish at trout farms. Over two weeks, two thousand rescuers removed 177,000 cubic feet of debris, pumped 73 million gallons of water, and pulled fourteen survivors from the wreckage. But seventy-five workers—those trapped in the turbine hall and in the flooded rooms below—weren't as lucky.

For Russians, the catastrophe called to mind the 1986 disaster at the Chernobyl Nuclear Power Plant in Ukraine, which was then part of the Soviet Union. Speaking on a Moscow radio station, Shoigu called the dam accident "the biggest man-made emergency situation [in] the past twenty-five years—for its scale of destruction and for the scale of losses it entails for our energy industry and our economy." Some commentators have called the events at Sayano-Shushenskaya the Russian Chernobyl. And just as Chernobyl raised questions globally about nuclear safety, Sayano-Shushenskaya has made other nations wonder: are other hydropower plants at risk?

Immediately after the accident, Russia's Federal Service for Ecological, Technological, and Nuclear

Supervision (Rostekhnadzor) launched an investigation. The official report, released on October 3, 2009, blamed poor management and technical flaws for the accident.

According to the report, repairs on Turbine 2 were conducted from January to March 2009, and a new automatic control system meant to slow or speed up the turbine in order to match output to fluctuations in power demand was installed. On March 16, the repaired turbine resumed operation. But it still didn't work right: the amplitude of the machine's vibrations increased to an unsafe level between April and July. The unit was taken offline until August 16, when the Bratsk fire forced managers at Sayano-Shushenskaya to press the turbine into service.

Back in operation, Turbine 2 vibrated at four times the maximum limit. As the control system decreased the turbine's output on the morning of August 17, the vibrations increased. The unit acted like the engine of an automobile being downshifted while speeding downhill, shuddering violently and stressing the fatigued metal pins holding it in place.

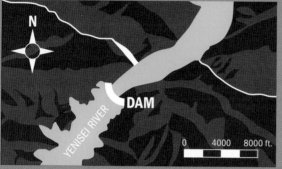

TIMELINE OF DISASTER

1. Fatigued by vibration, Turbine 2's fastening pins break at 8:13 AM. Water rushing down the penstock forces the 1,500-ton unit through the turbine hall floor and 50 feet into the air.

2. A geyser of water flowing at 67,600 gallons per second destroys the roof and floods the turbine hall. Power outages occur and communication systems fail.

3. The automated safety system also fails. Turbines 7 and 9 continue to operate even though they are submerged, causing short circuits, explosions, and structural damage.

4. Employees close the intake gates at the top of the dam at 9:30 AM, and the immediate crisis ends. In the following days, fourteen people are rescued from the debris; seventy-five lose their lives.

LMZ, the St. Petersburg metalworks that manufactured the plant's turbines, gave the units a thirty-year service life. Turbine 2's age on August 17 was twenty-nine years, ten months. Investigators determined that the power failure after the initial explosion had knocked out the safety system that should have shut down the plant, thereby turning a malfunction into a catastrophe.

The Aftermath

Officials from RusHydro and the Russian government have called for more stringent oversight of hydropower plants, but economic pressures may still put financial considerations ahead of safety. Six days before rescue efforts were halted on August 29, repairs at Sayano-Shushenskaya were already underway. Rebuilding will take five years and cost approximately $1.3 billion—but a pair of nearby aluminum smelters, property of global aluminum giant RusAl, can't wait that long. They consumed 70 percent of the damaged station's output and

need replacement power to maintain production. RusAl and RusHydro are pressing the government for additional financing to accelerate completion of a joint venture at Boguchansk on the Angara River, now in its twenty-ninth year of construction.

Russia's immediate solution to its power problem is to build more dams, but that won't fix a bureaucratic culture that seems to devalue safety. "If they were running a turbine with known deficiencies, they were putting economic concerns before human-life safety factors," says Eric Halpin, the special assistant for dam and levee safety for the U.S. Army Corps of Engineers. "If it's not safe, all other benefits—be they economic, environmental, or anything else—those all go away."

Twenty-first century living is, for many people throughout the world, safer and more comfortable than ever before. But sometimes the very things we rely on for such safety and comfort turn on us, leaving tragedy in their wake. Now that people have become used to amenities unheard of a century ago, they are loathe to give them up, no matter what the cost.

THE GULF OIL DISASTER

Deepwater Oil Rig Explosion on April 20, 2010 Leads to Environmental Catastrophe

What started as an epic drama in the Gulf of Mexico—beginning with a rare, lethal oil rig explosion, followed by a chain of never-before-attempted fixes to seal the underwater well—has finally settled into a sickening new reality. The best estimates in April, May, and June of 2010 had anywhere from 50,000 to 110,000 gallons of oil leaking into the gulf daily. With every failed attempt to plug the well, the world witnessed the next installment of what could be the greatest man-made environmental disaster in history. And there was no end in sight.

The incident began just after 7 PM on April 20, 2010 when an explosion rocked the offshore oil-drilling platform Deepwater Horizon, quickly engulfing it in flames. 41 miles off the coast of Louisiana, the platform was working for BP Exploration and Production, Inc., drilling in what is known as the Mississippi Canyon block 252. The first call for assistance to the Coast Guard went out at 7:23 PM, and within the hour, six helicopters were dispatched, soon followed by six surface cutters. Of the 126 men aboard the rig, 11 men were killed and 17 others were med-evaced to area hospitals with injuries.

While officials at BP and Transocean Ltd., the owners of the rig, did not say much at the time, industry insiders early on said evidence pointed to a blowout—meaning natural gas or oil unexpectedly flowed up the pipe, which almost always causes a catastrophic fire. In normal

Efforts to minimize damage caused by the oil spill in the Gulf of Mexico in the spring and summer of 2010 began immediately, as chemical dispersants were sprayed from airplanes onto the slick. British Petroleum also injected dispersants into the oil as it flowed from the well. Engineers were hoping that the chemicals, added to the oil, would cause it to form into droplets, which could be consumed by micro-organisms. The dispersants, however, are toxic themselves, but scientists believe they may be the lesser of two evils.

THE LEXICON OF A SPILL

The response to the Deepwater spill, and specifically the media coverage of it, provided a crash course in the basics of offshore drilling, adding top kill, heavy mud, junk shots and other seemingly self-explanatory terms to our collective lexicon. By the time anyone mentioned that the concept of hurling a slurry of golf balls, rope, and shredded tires into the spewing maw of the well head was not a practical joke, but standard practice for some above ground leaks, the entire endeavor began to feel slapped together and more than a little out of control in the spring and summer of 2010.

In fact, offshore oil production is a combination of blue-collar grit and a level of scientific rigor that rivals some NASA missions. Not that you'd guess that from the down-to-earth, but ultimately misleading terminology. Why, for example, were they using something as low-tech as mud to shore up what may be the worst oil spill in U.S. history?

The "Mud"

Before we get to the top kill, let's talk mud. The heavy mud that was pumped into the well was not something that was frantically dredged up from a nearby riverbank. Also called drilling mud, this highly-engineered material is used for everything from lubricating drill bits as they chew through subsea rock formations, to regulating the pressure in well bores to avoid precisely the kind of explosion that sank the Deepwater.

The "Kill"

In many cases, when drilling mud is "waded up" to offset a sudden increase in pressure coming from the subterranean formation, the mud is released deep in the well, through holes in the drill bit itself. When the well is being sealed completely, as opposed to using the mud to make slight pressure adjustments during drilling or pumping, it's called "killing" the well. The top kill is generally a less desirable version, where the kill happens from the top down, with mud forced into the kill lines built into the blowout preventers. However the mud is applied, from above or below, it is done slowly, and carefully.

The Straw

Remote-operated vehicles (ROVs) inserted a thin, 4-inch-wide pipe into the thicker, 6-inch wide riser that connects the rig to the subterranean well structure, hoping to suck the leaking oil to the surface. BP referred to this as a "riser insertion tool," but the media called it a "straw."

The Dome, the Tophat, and the Sombrero

Before the straw, there was the now-infamous dome, a 125-ton concrete-and-steel cap that was positioned over the open riser by robots. The hope was that the leaking oil and natural gas could be contained within the four-story dome, and drawn up through a tube to the surface. It took days for the container to be lowered to the ocean floor. It took far less time for the plan to fail, as the dome's umbilical hose quickly became clogged with methane hydrate, a slush that can form in oil wells when temperatures dip low enough.

BP also had a smaller container on the seafloor near the spill; it was delivered after the original dome was scrapped and could be deployed in the coming days or weeks. This "tophat" was fully equipped to fight methane hydrate formation and could be maneuvered into position relatively quickly. A similar steel container, called the "sombrero" because of its shape, was used more than 30 years ago to seal off an offshore oil leak in Mexico's Bay of Campeche. It took two tries to get the sombrero to work back in 1979.

The Junk Shot

If all else failed—the top kill, the tophat, or any number of jury-rigged solutions that engineers and researchers proposed when the well continued to gush—BP was ready to try the "junk shot." The name says it all: a mass of golf balls, rope and other seemingly random bits that were launched into the open riser. It was a tactic of last resort, and precisely as desperate as it sounds. And it too failed.

operations, the drill bit and string connecting the pipe to the surface bore a hole into and through various formations; because gas and oil (and water) exist under a higher pressure than the atmosphere, they are kept from flowing up by constant hydrostatic pressure from a fluid known as mud. As the well is drilled, it is lined with steel casing that is cemented in place.

During drilling, the pressures are constantly fluctuating and the mud must adjust to maintain hydrostatic pressure. Sudden kicks routinely happen. Large hydraulic valves called Blowout Preventers can both quickly seal off the wellhead and release gas, and usually the level of mud is adjusted and drilling can resume.

Let's add to the mix. The work BP was and still is doing in the Gulf of Mexico is drilling in what's known as deep and ultra-deep water—from 5,000 to 10,000 feet of water, and up to another 30,000 feet underground—and it's riskier and far more complex than drilling in shallow water or on land. The deeper the well, the higher the temperature and pressure of the fluids or gas tapped, and as those fluids rise, they expand quickly. In this high-pressure, high-temperature drilling, even small gas bubbles become enormous as they travel toward the surface. "That's the inherent danger of drilling deeper," said Bill Markus, vice president of response at the legendary Texas firm of Boots & Coots. "You have small gas influxes all the time, but in ultra deep they become radical."

In addition, the farther offshore the rig, the more complex each piece of the logistical puzzle becomes, especially when tragedy strikes. Fires on any drilling rig are rare, but when they happen, they are very dangerous.

No oil company wants a blowout and they spend vast sums of money and expertise to make sure the chances are slim. But blowouts do happen. While

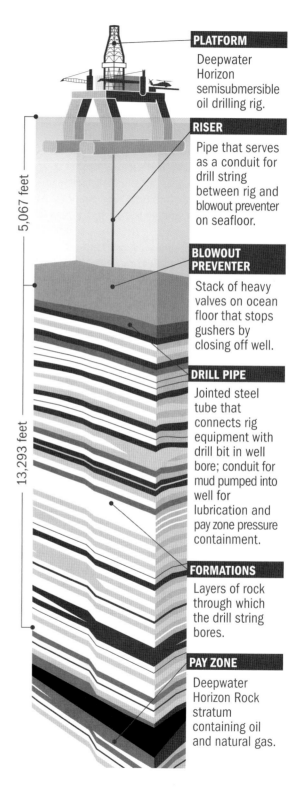

PLATFORM
Deepwater Horizon semisubmersible oil drilling rig.

RISER
Pipe that serves as a conduit for drill string between rig and blowout preventer on seafloor.

BLOWOUT PREVENTER
Stack of heavy valves on ocean floor that stops gushers by closing off well.

DRILL PIPE
Jointed steel tube that connects rig equipment with drill bit in well bore; conduit for mud pumped into well for lubrication and pay zone pressure containment.

FORMATIONS
Layers of rock through which the drill string bores.

PAY ZONE
Deepwater Horizon Rock stratum containing oil and natural gas.

5,067 feet

13,293 feet

"Blowouts happen four to five times a year."

most insurance companies attribute them to human error, each event is unique. "If this was a blowout," Markus says, "it's the first deep-water one in the Gulf of Mexico that I can think of. And for that to happen requires the perfect storm of scenarios."

The Deepwater Horizon was a state-of-the-art, dynamically positioned floating semi-submersible drilling platform that cost BP $502,000 a day to run. On April 20, it was close to finishing drilling a series of exploratory wells in 5000 feet of water and was scheduled to be moved. A report obtained from a drilling industry analyst who asked to remain anonymous (due to a pending investigation) shows that rig workers had recently set and cemented a casing and were in the process of displacing the riser (a pipe that encloses the drill string from the platform to the sea bottom) with seawater and setting a surface plug when the well blew out.

The report reads: "The incident appears as if it was either the product of gas migration through the cement sheath as it was setting or that the act of displacing the riser to seawater reduced the hydrostatic head enough that it caused the well to start flowing though cement that was not yet hard. With casing in the ground, these two possible explanations seem by far the most likely, although its impossible at this point not to rule out some other cause or contributing factor such as a blowout preventer or a casing integrity issue."

"Blowouts happen four to five times a year," said Markus, "down from 11 to 15 in 2000."

Two months after the blowout, in July of 2010, the best-case scenario hoped for by British Petroleum was that a pair of relief wells would intersect with the damaged one, and the company would be able to fill it with concrete. Meanwhile, the public watching the spill in the media—and who saw the aftermath on the beaches and fishing waters—was growing more and more skeptical that any of the proposed solutions would work. Another possible outcome involved the disintegration of the well head or other structural elements due to the efforts to divert or block the flow. If that happened, exponentially more oil would gush into the gulf. The public began to realize that it could get much worse before any improvements were seen.

For the communities and wildlife already hit by the first incursions of oil, the fight to stem the flow was a losing battle. While teams of million-dollar underwater robots worked at the well's rupture on the ocean floor, the spill response topside could hardly get more low-tech. Skimming vessels attempted to vacuum up slicks, but a gallon of oil can spread to the size of a football field in an hour, quickly becoming too thin for skimmers to be effective. The spill had also plunged below the miles of boom floating in the gulf. So the stricken marshes were being dabbed, by hand, with special cloths and paper towels.

There are far more innovative solutions to spill cleanup, particularly in the area of advanced materials that selectively absorb oil without soaking up water. In 2008, researchers at MIT unveiled a mesh of nanowires that formed a kind of paper. It could stay dry in water for months while soaking up oil and other hydrophobic liquids. The mesh could then be heated to release and recover the captured oil, and the nanopaper redeployed to the spill.

That a breakthrough material discovered in the lab two years prior wouldn't be ready for use in the Deepwater Horizon spill is no surprise. But there's another advanced material to target oil that has been available for more than a decade. AbTech CEO Glenn Rink founded his Arizona-based

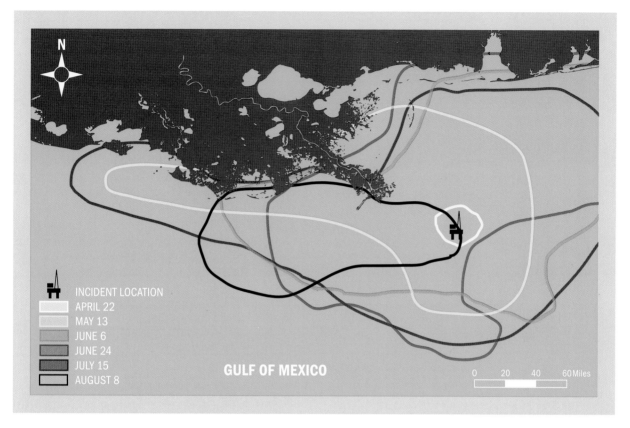

N

INCIDENT LOCATION
APRIL 22
MAY 13
JUNE 6
JUNE 24
JULY 15
AUGUST 8

GULF OF MEXICO

0   20   40   60 Miles

company as a direct response to the havoc wreaked by the Exxon *Valdez* and other, less-publicized, tanker spills. In 1997, after years of testing that included a limited deployment in a spill in Aruba, AbTech launched the Smart Sponge. The solid material floats on the surface of the water, either incorporated into lengths of boom or dropped into specific areas from boats or aircraft. Like MIT's mesh, it attracts and captures oil and chemically similar pollutants while ignoring water, and can be left in place for months. But the Smart Sponge permanently bonds the oil, and the only way to access that stored energy is to burn the material itself. "It was a closed loop. We saw that as being really valuable," Rink said. But so far, no one has put an order in for the sponges.

The fight to stem the flow [of oil] was a losing battle.

The biggest reason for the lack of interest in the Smart Sponge, according to Rink, was the strange business of oil-spill cleanup. Oil firms don't typically deploy their own skimmers or towel-wielding employees—they hire someone else to do it. An entire industry of spill-response companies serves the oil industry at large, and in many cases a single outfit will be contracted to respond to potential spills from multiple energy firms. And for the most part, these spill responders are paid by the hour, using materials that are less efficient than the Smart Sponge but cheaper and more easily processed to recover oil.

Since BP and other energy firms aren't directly involved with cleaning up their spills, the Smart Sponge fell into a logistical limbo. Spill responders would need to buy and store tons of the polymer-based material, which would raise their overall fees while in return bill fewer hours. Rink and AbTech were forced to move on, applying the material to a range of water-filtration applications. Not long before the Gulf Oil Spill, *Time* magazine featured

The Deepwater Horizon oil spill in the spring and summer of 2010 was the largest accidental marine oil spill in the history of the petroleum industry. It was estimated that more than 50,000 barrels per day spilled from a sea-floor oil gusher after the Deepwater Horizon drilling rig exploded on April 20, 2010. The leak was stopped when engineers capped the gushing wellhead on July 15, and on September 19, a relief well was completed, rendering the well "effectively dead." Officials believe that nearly 5 million barrels of oil were ultimately released into the Gulf of Mexico.

Rink on a list of entrepreneurial heroes. Yet the oil industry barely noticed.

The Deepwater Horizon debacle could change that. "We've been getting inquiries, including from BP, about how much product we have on hand, how quickly we can deploy it," Rink said. He appeared on *Glenn Beck*, and a once-forgotten solution is suddenly found featured in online blogs and newspaper articles. Whether or not the oil industry can reconfigure itself to make large inventories of readily deployable, oil-targeting materials feasible, the demand is there. "There are no active purchases," Rink said, "but we're told orders are imminent. We've priced out probably over $10 million in sales."

The Aftermath

By August, after a summer of setbacks and failed top kills designed to stem the flow of oil into the Gulf, a "static kill" that cemented the well from the top down was finally successful, and the casing was filled. The next task at hand was to drill a relief well, which cemented the well from the bottom up so that all efforts could turn to recovery operations. Unfortunately, oil in the Gulf and on the shores continues to be a major problem, and it will be years before scientists will be able to fully understand the extent of the damage caused by the Deepwater catastrophe.

DEEPWATER HORIZON SPILL BY THE NUMBERS

The Leak

Barrels of oil leaked	4.9 million
	(205.8 million gallons)
Times more oil leaked than Exxon *Valdez*	19
Barrels leaking per day when well first broke	62.000
Barrels leaking per day after capping on 7/15/2010	53,000
Dollars' worth of oil spilled at current market price ($81.17 per barrel)	397.7
Miles of coastline contaminated by oil	665

The Cleanup

Gallons of oil chemically dispersed by National Incident Command	16.5 million
Gallons of oil naturally dispersed into droplets smaller than the diameter of a human hair	32.9 million
Gallons of oil evaporated or dissolved. Instead of breaking down into small droplets, the oil breaks apart molecularly and dissolves into the water.	51.5 million
Gallons of oil skimmed off the Gulf by the more than 830 skimming vessels used in the response	6.2 million
Gallons directly recovered from the wellhead into ships through the riser pipe and top-hat systems	35 million

Gallons of oil removed by a series of 411 controlled burns	11.4 million
Gallons of oil still remaining in the water or washed ashore	53.5 million
Potential gallons of gasoline leaked from wellhead (about one-fourth of daily consumption in the United States).	95.6 million the
Feet of sorbent boom (8.7 million) and containment boom (2.7 million) currently deployed to contain the oil	10.4 million
Total number of personnel currently deployed in response to the spill. On July 8, 47,000 people had been deployed.	28,900
Square miles of Gulf waters that remain closed to fishing	57,539

In Perspective

Times you could drive a Toyota Prius (48 mpg highway) around the Earth at the equator using the lost oil	184,181
Times you could drive a Hummer H3 (18 mpg highway) around the Earth	69,068
Olympic-size swimming pools that could be filled with the oil that leaked from Deepwater Horizon	311
Homes that could have been heated for one year	13,208

THE BIGGEST ENVIRONMENTAL DISASTERS IN THE WORLD

Chernobyl Power Plant

The damage caused by the catastrophic nuclear meltdown in Chernobyl, Ukraine, is difficult to measure. There were more than 50 direct fatalities on April 26, 1986, but thousands of cancer deaths can be attributed to direct exposure to radiation, and thousands more suffer with brain tumors, massive headaches, and birth defects. The city of Chernobyl is a ghost town now, with radiation levels not fit for habitation.

Abandoned gas masks lay on the floor of a classroom in a school in the deserted town of Pripyat, adjacent to the Chernobyl nuclear site in May 2003. Pripyat, which had 45,000 residents, was totally evacuated in the first three days after the reactor number four at the Chernobyl plant blew up on April 26, 1986, spewing out a radioactive cloud and contaminating much of Europe.

Three Mile Island

The Three Mile Island nuclear meltdown began on March 28, 1979, and consisted of three days of mechanical and human failures that led to a crisis that captured the world's attention. Despite the fact that there was no significant amount of radiation released in the catastrophe, there was nearly a billion dollars in cleanup costs, and the incident brought a public awareness to nuclear power plants.

Bhopal Gas Leak

On December 3, 1984, more than half a million residents of Bhopal, India, were poisoned when deadly toxins were released from the Union-Carbide pesticide manufacturing plant. The death toll stands at approximately 35,000, but countless survivors continue to suffer congenital defects and serious health issues. The site has yet to be cleaned up, and Dow Chemical Company, which owns Union-Carbide, has mostly eluded any charges of negligence.

Exxon *Valdez* Oil Spill

It wasn't the largest oil spill in history, but when the Exxon *Valdez* tanker struck an iceberg in Alaska on March 23, 1989, images of the environmental damage were forever seared in the public's memory. Whales washed up on shore and dead birds covered in oil at Prince William Sound became symbols of the havoc man is capable of causing in nature. Thousands of workers spent years cleaning up the damage, but the beaches may never be the same.

Love Canal

Love Canal, New York, was a small neighborhood just outside Niagara Falls, with residents who had no idea they had purchased homes on top of what used to be a major toxic waste site. By the 1970s, when the rate

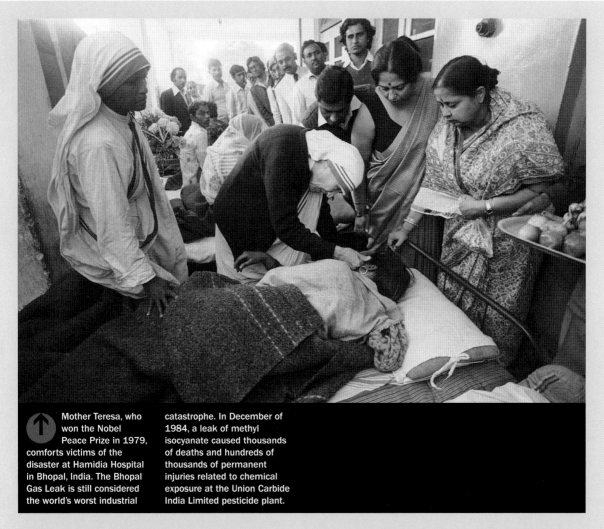

Mother Teresa, who won the Nobel Peace Prize in 1979, comforts victims of the disaster at Hamidia Hospital in Bhopal, India. The Bhopal Gas Leak is still considered the world's worst industrial catastrophe. In December of 1984, a leak of methyl isocyanate caused thousands of deaths and hundreds of thousands of permanent injuries related to chemical exposure at the Union Carbide India Limited pesticide plant.

of miscarriage, birth defects, and cancer diagnoses in Love Canal was far above normal population rates, the U.S. declared a federal health emergency and demolished hundreds of homes, relocating families. Today, the site is considered clean, but the neighborhood has not returned to what it once was.

Minimata Mercury Incident

When the Chisso Corporation in Japan began dumping thousands of tons of industrial waste in the Hyakken Harbor and waters surrounding the Chisso Factory from 1908 to 1968, thousands of residents in the area reported symptoms of mercury poisoning. Nearly 2,000 people have died, and thousands more were stricken with paralysis, fell into comas, or suffered insanity. The company spent millions compensating victims and has been ordered to clean up the contaminated sites.

TVA Coal Spill

On December 22, 2009, more than 5 million cubic yards of toxic coal sludge exploded over a dam wall at the Kingston Tennessee Valley Authority power plant, contaminating the Emory River and hundreds of acres of farmland in the area. The coal ash contained arsenic and other carcinogenics, and environmental experts have conceded that it will be impossible to do a complete cleanup from the catastrophe.

SURVIVAL TIPS:
SURVIVING AN EXPLOSION

In an explosion, inches and seconds can often mean the difference between life and death. Quick thinking, quick reflexes, and awareness can save lives.

➲ Learn to recognize the signs of an impending explosion. The strong, pungent smell of sulfur often precedes a chemical ignition. If you see, hear, or smell anything out of the ordinary, get away.

➲ Cover up. Explosions can send chunks of metal, stone, or concrete flying at breakneck speeds, and you don't want to be in the way. Find something solid to hide behind, or get down on your stomach if you're caught in the open.

➲ Get help. Burns and shrapnel wounds can kill. Seek immediate medical attention for yourself and others as soon as possible.

← The remnants of an abandoned class-room are seen in a pre-school in the deserted town of Pripyat on January 25, 2006 in Chernobyl, Ukraine. Pripyat and the surrounding area will not be safe for human habitation for several centuries. Scientists estimate that the most dangerous radioactive elements will take up to 900 years to decay sufficiently to render the area safe.

Dozens of vehicles sit in the northbound lanes of Interstate 75 after a deadly pileup along a foggy stretch of road on March 14, 2002 in Ringgold, Georgia. Four people were killed in the chain-reaction accident involving around 125 vehicles on both directions of the busy interstate. Officials were astonished that there were not more fatalities, given the carnage on the scene.

TRANSPORTATION

Mobility made America. Before trains, travel was limited by the speed of a horse. Crossing the country was a long arduous journey—and many graves dotted the roadside as a result. Train travel made it possible to traverse long distances quickly and in relative safety. Automobiles soon followed. No longer were travelers confined to fixed pathways; they could go anywhere there was a road. But although trains and cars eliminated many of the obstacles of long-distance travel, they also created new and different dilemmas.

The infrastructure in the U.S. needs work. Americans are driving twice as many miles as they did twenty-five years ago, while the number of miles of new highway has increased by less than 4 percent in that period. One-quarter of the 599,893 bridges in the United States have structural problems or outdated designs.

However, driver miscalculations still play a big role. The speeds at which cars, trains, and trucks travel today, combined with their sheer mass, can result in catastrophe. And although carmakers are now creating collision warning systems, easing up on the gas pedal could prove to be an immediate, low-tech solution that saves time, money, and, ultimately, lives.

MIANUS RIVER BRIDGE

Greenwich, Connecticut's Surprise: June 28, 1983

At 1:30 AM on June 28, 1983, a 100-foot-long suspended span, part of the bridge carrying Interstate 95 over the Mianus River in Greenwich, Connecticut, collapsed, throwing two trucks and two automobiles into the water. Three people died and three were seriously injured.

The bridge was of a common construction. The suspended span had been attached to the bridge structure by a pin-and-hanger assembly at each corner. Each assembly contained two pins—an upper pin attached to the anchored, shore-side portion of the bridge, and a lower pin attached to the suspended span. A pair of steel hangers, 1.5 inches thick, connected each set of pins.

When the cap of the lower pin on the inboard side was removed during the postmortem inspection, black rust flowed out. Said one investigator, "[T]here is no way that I would have ever believed that there could have been that amount of deterioration behind that pin cap."

There are more than half a million bridges in the United States, and some experts estimate that one-quarter of them have structural problems or outdated designs. Help may come from the use of a new portable, computerized, nondestructive testing system developed by metallurgist Edward Escalante and materials researcher Eric Whitenton of the National Bureau of Standards (now the National Institute of Standards and Technology). The system uses two probes: one to polarize steel and the other to measure voltage changes. Steel with little or no corrosion shows a high polarization resistance, and vice versa.

↑ Wreckage from the Mianus River Bridge collapse on Interstate 95 in the Cos Cob section of Greenwich, Connecticut, after a 100-foot section of the northbound span collapsed on June 28, 1983. Three people were killed when their cars plunged some 70 feet below. Casualities would have been greater had the accident not occurred at 1:30 AM when traffic was light.

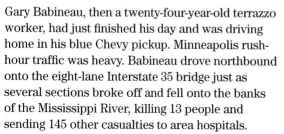

Gary Babineau, then a twenty-four-year-old terrazzo worker, had just finished his day and was driving home in his blue Chevy pickup. Minneapolis rush-hour traffic was heavy. Babineau drove northbound onto the eight-lane Interstate 35 bridge just as several sections broke off and fell onto the banks of the Mississippi River, killing 13 people and sending 145 other casualties to area hospitals.

"It didn't even make a noise before it collapsed," Babineau recalls. "It just fell away." Part of the bridge split in two right under his truck, sending it into a 35-foot nosedive as it chased the road deck straight to the ground. Babineau's truck landed on the edge of the fractured roadway, bending its frame at a 20-degree angle like a fortune cookie. "I thought my back had snapped in two," Babineau said.

As he reached for his door handle, Babineau could hear the wild screeching of tires right above him, where what was left of the bridge now hung precipitously. Then one car skidded over the edge and crashed next to Babineau's pickup. He balled himself up and stayed as low as possible in his cab. Then came another screech and another crash, then a third. Then silence. He opened his door and looked up to see a white sedan that had managed to stop before landing on him. It had been ten, maybe fifteen seconds since the collapse.

Babineau approached a woman who was in a car in front of him. They were both stuck on the collapsed section of bridge. They had made it to the side of the bridge and were climbing off when Babineau heard children screaming and crying. He ran up to the bridge and climbed up to a bus, working with a few other survivors to help more than fifty children—many of whom were injured—get off the bus, off the bridge, and onto safe ground. It took just three minutes to evacuate the children.

The Aftermath

The rescue effort was massive and immediate, involving city, county, and state rescue workers, the U.S. Army Corps of Engineers, the U.S. Coast Guard, and scores of civilian volunteers. Ninety-three people were rescued in three hours, but the recovery of submerged vehicles and bodies took about three weeks.

"It didn't even make a noise before it collapsed...."

The aftermath of the I-35W Mississippi River bridge collapse in Minneapolis, Minnesota, on August 1, 2007. 13 people were killed and 145 were injured when Minnesota's fifth busiest bridge plunged to the river and riverbanks below during evening rush hour. The National Transportation Safety Board concluded that design flaws of the gusset plates were responsible for the bridge's failure. Remarkably, 93 people were rescued, as city, county and state rescue workers arrived on the scene shortly after the collapse. However, it did take several weeks to recover the bodies and vehicles that were submerged in the Mississippi River.

MINNEAPOLIS BRIDGE COLLAPSE

On August 1, 2007, the I-35W Bridge over the Mississippi River in Minneapolis, Minnesota, suffered a failure in the river span of the deck truss, which caused a complete collapse of the entire truss structure as well as some of the approach spans. Thirteen people were killed and one hundred were injured in the bridge's unexpected collapse.

At just after 6:00 PM during midweek rush hour, traffic along the bridge was moving slowly when the central span of the bridge dropped first, quickly followed by adjoining spans, which collapsed into the river and riverbanks below. The south part of the bridge dropped more than 80 feet to the water, while the north section fell into a railyard and onto several freight train cars. A few of the automobiles on the bridge became submerged in the water, while others caught fire. Sixty children on board a school bus were fortunate to escape as the bus nearly toppled into the river.

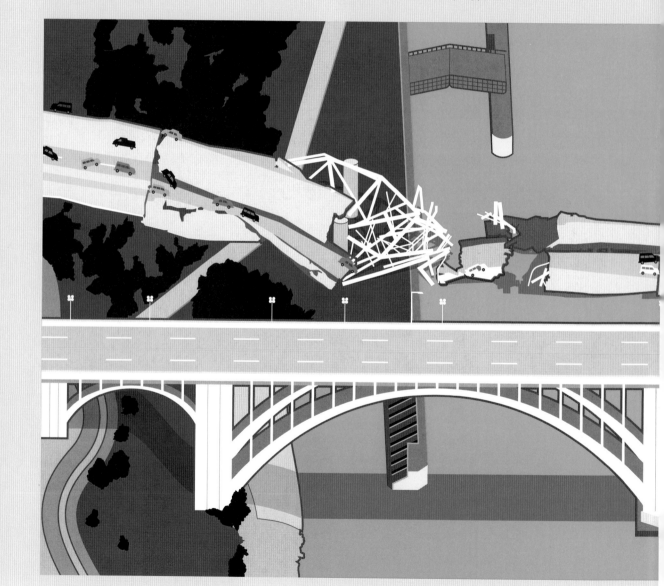

The National Transportation Safety Board began a comprehensive investigation and just five months later, they announced that they had determined that a design flaw was responsible. Sixteen fractured gusset plates in the center span on Interstate 35W were a main cause of the deadly bridge collapse, according to the NTSB.

The plates, which connected steel beams in the truss bridge, were roughly half the thickness they should have been because of a design error. Extra weight from construction was also a factor in the tragedy. The findings confirmed forecasts by investigators from three months after the collapse–and by engineering experts in the immediate aftermath–and underscored the dire state of America's crumbling infrastructure.

According to the NTSB photos, gusset plates were buckled more than two years before the accident. But there were many other factors that contributed to the collapse. Poor maintenance, corrosion, older welding technology and fatigue cracks in combination with Minnesota's harsh winters and application of de-icing materials only added to the catastrophic destruction of the I-35W bridge in Minneapolis.

INFERNO ON INTERSTATE 5

Los Angeles and Long Beach, California: October 12, 2007

It was just after 10:45 PM when Los Angeles County fire inspector Ron Haralson received a call alerting him to a multivehicle pileup in the southbound tunnel of a truck bypass on I-5 about 30 miles north of downtown Los Angeles. Fifteen minutes later, he saw the flames as he sped north on the state's major north-south interstate, with the narrow, two-lane truck tunnel crossing below. "It was all orange and all fire," he recalled, "like a blowtorch out of both ends."

When Haralson pulled up to the tunnel, several dozen firefighters were already there. "We got word that there were other vehicles in there, but we just couldn't get to them," Haralson said. It took three hundred firefighters nearly twenty-four hours to fully extinguish the flames. In that time, almost thirty big-rig trucks and one passenger vehicle burned to steel skeletons in a 1,400-degree inferno that killed three, shut down California's central artery for more than two days, and cost $17 million in cleanup and repairs. Interviews with experts and first responders revealed that several elements common to many large chain-reaction vehicle pileups contributed to the accident: poor road conditions, low visibility, and heavy traffic.

The first factor was the rain that had begun falling just before the crash—the region's first significant precipitation in months. Since oil tends to build up on roads during dry spells, a sudden rain creates exceptionally slick conditions. At about 10:30 PM, a tractor-trailer hauling coffee beans exited the tunnel onto the wet roadway. The driver lost control and careened a full third of a mile before crashing into a guardrail. Remarkably, the truck didn't hit any other vehicles, but the debris left in the road most likely caused slowdowns that contributed to a two-truck collision near the tunnel's exit minutes later.

Low visibility also played a part. The blind curve inside the 550-foot tunnel makes it impossible for drivers at the entrance to see through to the other end and brake in time to avoid becoming part of an existing pileup. "I came to a stop, and then they just started hitting me, one right after another," truck driver Tony Brazil told a local TV news station shortly after the crash.

Traffic was also a factor. The nearby port complex of Los Angeles and Long Beach is the fifth busiest in the world, receiving 40 percent of all goods shipped into the United States. Two-thirds of those goods leave Los Angeles and Long Beach by truck, which is why it's not surprising to see superhighways like the I-5 packed with traffic at all hours.

Overall, Americans are driving twice as many miles as they did twenty-five years ago, while the number of miles of new highway has increased by less than 4 percent in that period. Traffic density is a primary risk factor for chain-reaction crashes, and adding big rigs to the mix raises the risks. Eighty-three percent of all fatal truck accidents are multivehicle crashes, compared to 61 percent of all fatal passenger vehicle crashes. "It's simply a matter of physics," says Daniel Blower, who researches truck safety at the University of Michigan. "Because trucks are so much bigger and stiffer, they keep going—and many more vehicles may therefore be involved in a crash."

The Aftermath

The night of the accident on Interstate 5, about twenty people escaped from their vehicles and ran out of the tunnel before it was engulfed in flames. Two truck drivers, Ricardo Cibrian and Hugo Rodriguez, were found dead in their vehicles near the tunnel's exit, as was Rodriguez's six-year-old son, Isaiah, who had gone to work with his father

"It was all orange and all fire...."

Two of many trucks burn in the truck lanes of the southbound Interstate 5 in the Newhall Pass between Santa Clarita and Los Angeles, California, early Saturday morning, October 13, 2007. The multivehicle pileup was caused by poor road conditions, low visibility and heavy traffic as the first rain in months began to slick the pavement. The pileup killed two truck drivers and a six year old boy. Highway crews worked around the clock for a month to repair the tunnel, and a new lighting system was installed to help improve visibility. But officials point to a newly reduced speed limit as being instrumental in avoiding another I-5 catastrophe.

that day. Ten other drivers suffered mild to moderate injuries.

It took highway crews a month to repair the tunnel. A new lighting system was installed, and the walls were painted white for better visibility. Perhaps the most important renovation is the interactive sign that posts the new speed limit: 45 mph, down from 55 mph.

Unfortunately, the conditions that led to the I-5 catastrophe are repeated daily on highways throughout the country. Although carmakers are developing radar-based forward collision warning systems, easing up on the gas pedal could prove to be an immediate, low-tech solution that saves time, money, and ultimately, lives.

When people think of transportation disasters, they usually think of plane crashes and car accidents. Train derailments seem like a thing of the past, more suited to the Civil War and the Wild West than the twenty-first century. But train accidents still happen. The speeds at which trains travel today, combined with their sheer mass, can result in twisted hunks of metal—and that's only the beginning of the aftermath of such mishaps.

It was a slow day on Norfolk Southern's R Line, a lonely 82-mile section of track that winds through rural South Carolina. So when the crew of a local train in Graniteville, a small town just northeast of the Georgia border, radioed in for clearance to make deliveries along the line, the dispatcher quickly gave his OK. In radio parlance, he issued a track warrant.

The three-man crew finished up around 7 PM and decided to park the train overnight on the short spur track that leads to the Avondale Mills textile plant. Using a special key, one crewman unlocked the switch and manually repositioned it to divert the train onto the sidetrack. The crew then radioed the dispatcher more than 100 miles away and cleared the track warrant. The R Line reopened to through trains.

Seven hours later, at around 2 AM on January 6, Norfolk Southern Train 192 pulled out of Augusta, Georgia, on the R Line, bound for Columbia, South Carolina. The train's two locomotives were pulling forty-two cars that morning, among them three tankers loaded with 270 tons of liquefied compressed chlorine gas. As the train neared Graniteville, the engineer, twenty-eight-year-old Christopher Seeling, would have been routinely scanning the track ahead.

What he saw must have made his blood freeze. Instead of the normal white reflector at the spur track junction, there was a red reflector. He frantically applied the emergency brake, but it was too late. Seventeen seconds later, Train 192 veered onto the spur track and slammed into the parked local train.

Amid a crescendo of tearing metal, the locomotives and the first fourteen cars of Train 192 derailed, splaying out like pick-up sticks. Two of the tankers carrying chlorine remained upright and intact. The third catapulted onto its side, and the impact punched a 29-inch gash in its midsection. The liquefied chlorine began to spew out, vaporizing to form a huge greenish-yellow cloud of toxic gas that soon drifted over the town.

By the time the fumes dissipated, 9 people had died. Another 550 sought medical treatment, and more than 5,400 Graniteville residents were told to evacuate. It was the worst U.S. rail accident involving hazardous materials in twenty-six years.

The National Transportation Safety Board found that the switch was in the wrong position and its report blamed the accident on the P22 train crew's failure to reline it for mainline operations. The company fired the three crewmen responsible. The Federal Railroad Administration requires crews to return switches to the normal position before they clear track warrants, but not to explicitly report switch positions. Some rail companies do require this, but at the time of the accident Norfolk Southern had no such policy; four days later, after the accident, it implemented one.

Electronic signals have been used for decades to alert dispatchers of switches' positions. But

The liquid chlorine began to spew out, vaporizing to form a huge greenish-yellow cloud.

Two Norfolk Southern trains collided near an Avondale Mills Plant in Graniteville, South Carolina, on January 6, 2005. Train No. 192 was carrying chlorine gas, sodium hydroxide and resol when it was improperly diverted to a siding and crashed into a parked train No. P22. Nine people were killed in the accident, and more than 250 were treated for exposure to chlorine, as officials were forced to evacuate the area for two weeks to decontaminate.

about one-third of Norfolk Southern's 16,630 miles of track are unsignaled, as are about 40 percent of the 170,000 miles of mainline tracks in the country. According to the railroad industry, it would cost $100,000 per mile to modernize this "dark territory."

Yet another threat looms on U.S. rails. The pressure tank cars carrying chlorine on Train 192 were made from relatively safe heat-treated steel. But about 60 percent of the nation's sixty thousand pressure tank cars were built using ductile steel—brittle, untreated metal that was the standard before 1989. These cars are permitted to remain in service for fifty years after their manufacture, so some will be in use until 2038.

The Aftermath

On January 31, 2005, another Norfolk Southern train derailed in Pennsylvania, spilling anhydrous hydrogen fluoride; nearly two hundred people were evacuated. In response to these and other rail accidents, a bipartisan bill was introduced in the Senate to toughen federal oversight of rail safety.

In 2002, 3 million tons of chlorine—about 37,000 tank carloads—were moved by rail in North America. Unless the railroad industry upgrades tank cars and lines, the next Graniteville disaster could be much worse.

RUNAWAY TRAIN

On January 6, 2005, a railroad switch set in the wrong position caused a through train going 45 mph to barrel down a sidetrack near Graniteville, South Carolina. It collided with a parked locomotive. The perspective shown here is from just north of the accident.

1. THE IMPACT: Norfolk Southern Train 192, headed for Columbia, South Carolina, plowed into a local train parked overnight on a spur track.

2. THE CHEMICAL: Tank cars Nos. 6, 7, and 9 were carrying liquefied compressed chlorine gas. All three derailed.

3. THE VICTIMS: Four hundred workers fled the Avondale Mills textile plant. Six died before they could escape the toxic fumes.

4. THE CLOUD: The evacuation radius extended a mile from the crash site. Some residents didn't return for weeks.

5. THE CAUSE: A federal investigation later revealed that the railroad switch had been "lined and locked" for the sidetrack instead of the main track.

THE OLD STANDARD

Hand-thrown switches have been around since the early 1800s.

WHERE THEY ARE: Thousands are still used on more than 60,000 miles of mainline track.

HOW THEY WORK: Once unlocked, the lever is depressed by foot. This turns the reflector 180 degrees, changing its color and the direction of the track.

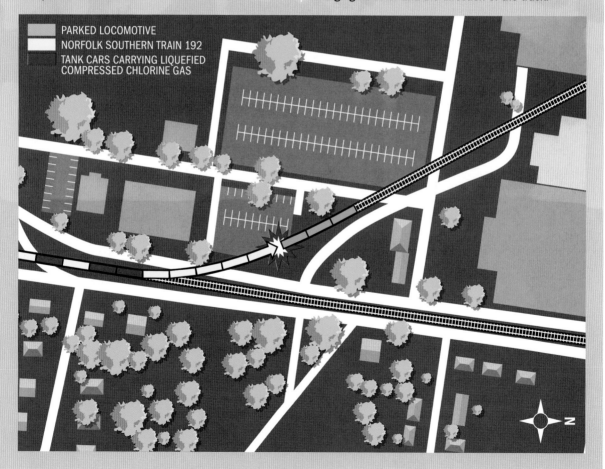

PARKED LOCOMOTIVE
NORFOLK SOUTHERN TRAIN 192
TANK CARS CARRYING LIQUEFIED COMPRESSED CHLORINE GAS

Egyptian rail road workers stand beside a train which caught fire about 80 kilometers (50 miles) south of Cairo on February 20, 2002.

Wrecked vehicles burn after a train crashed and burst into flames on the outskirts of the Iranian city of Mashhad on February 18, 2004. Close to 200 people were killed and hundreds injured when runaway rail wagons loaded with sulphur, petrol, and fertiliser derailed and blew up in northeast Iran.

10 LARGEST GLOBAL TRAIN DISASTERS

Location	Date	Deaths
Ural Mountains, Russia	1989	575
Modane, France	1917	550
Salerno, Italy	1944	521
Ayyat, Egypt	2002	361
Firozabad, India	1995	358
Neishabour, Iran	2004	320
Virilla River Canyon, Costa Rica	1926	300
Calcutta, India	1999	285
Mansi, India	1981	268
Buenos Aires	1970	236

SURVIVAL TIPS: CAR PILEUPS

After the Interstate 35 bridge collapse, Gary Babineau avoided a classic mistake by staying in his truck until the crashing around him stopped.

➲ "It's instinctual to get out of your car," says Sergeant Doug Sheets, of the California Highway Patrol. "But that's how people get killed. You can take a pretty big hit in a car if you're wearing a seatbelt."

➲ Sheets suggests rolling down your window before making any moves. "You can hear accidents unfolding," he said. All quiet? Then make a break for it. "Even if your car has four flat tires, drive out of the way," Sheets adds. "It's easier for investigators if you don't, but nobody's going to blame you if it saves your life."

➲ America's roads are dangerous, and drivers take their lives into their hands every time they turn the key in the ignition and back out of the driveway. But bridge collapses are the exception, not the norm: most accidents are caused by simple human error.

INDEX

PHOTOGRAPHY CREDITS

AlaskaStock.com: Randy Brandon: 190

Richard Armstrong: 62

AP Photo: 42, 81, 143 right, 144 left, 150, 225; Stacy Bengs/ The Minnesota Daily: 227; Rod Boren: 145 left; Jeff Gentner: 205; Bullit Marquez: 39; Stephen Morton: 199; Robert Nichols: 142; Dennis Oda/Star Bulletin: 145 right; Eraldo Peres: 141; Fabio Pozzebon: 137; Jack Thornell: 100, 103

Barnard-Stockbridge Historic Photograph Collection, Special Collections & Archives, University of Idaho Library, Moscow: 50

Stephen Brooks: 153

City of Toronto Archives: 183

Compiled by Climate Change Information Centre: 126-127

Brett Coomer: 200

Corbis: Matiullah Achakzai/epa: 114; Arctic-Images: 30; Richard Baker/In Pictures: 194; Kapoor Baldev/Sygma: 219; Beawiharta/Reuters: 131; Bettmann: 86, 89, 97, 124, 143 left; Gene Blevins/Los Angeles Daily News: 231; Geraldo Caso/epa: 12; DPA/epa: 236; Marcelo Hernandez/epa: 13; Hulton-Deutsch Collection: 176, 179 top; F.C. Koziol: 71; The Mariners' Museum: 185; Sean Masterson/epa: 61; Eric Nguyen: 78, 83, 92; David J. Phillip/epa: 105; Alberto Pizzoli/ Sygma: 132-133; Steve Pope/epa: 119; Reuters: 149; Ed Sackett/Dallas Morning News: 144 right; Svenja-Foto: 146 left; Marilynn Young/Orange County Register: 58

Getty Images: 146 right, 151, 220, 235 bottom; AFP: 72, 218; Fred Dufour/AFP: 4; Natalie Fobes: 191; Forca Aerea Brasileira via Latin Content: 140; Nicholas Kamm/AFP: 10; Tim Laman: 68; Erik S. Lesser: 222; Flip Nicklin: 6; William Philpott/AFP: 203; Joe Raedle: 118; Justin Sullivan: 55

Garrett Grove: 76

Chad Hunt: 56
Library of Congress: 19, 20, 48, 51

Mary Evans Picture Library: Illustrated London News Ltd: 117

NASA: 40, 94, 108, 158, 161, 165, 169, 170, 174; Steven R. Nagel: 167; Jeff Schmaltz/MODIS Rapid Response Team/ Goddard Space Flight Center: 53; Dryden Carla Thomas: 172

National Geographic Stock: Mark Thiessen: 44

NOAA Photo Library: 98, 128; Lieutenant Commander Mark Moran, NOAA Corps, NMAO/AOC: 111

Nova Scotia Archives and Records Management: 179 bottom; Gauvin & Gentzel: 182

Don Poleto: 123

Redux: Damon Winter/The New York Times: 14

Reuters: 134, 147 left, 147 right; Aladin Abdel Naby: 235 top; Ilya Naymushin: 196, 207; Eric Thayer: 156

Vyto Starinskas: 121

Andrew Tuthill/ U.S. Army Cold Regions Research and Engineering Laboratory: 122

U.S. Coast Guard: 187, 188, 189, 216

Courtesy of the USFS Region One Archives: 47

U.S. Geological Survey: 8-9, 24, 28, 33; D. Dzurisin: 35; U.S. Army: 25, 26

U.S. Navy: Photographer's Mate Airman Jeremy L. Grisham: 112; Mass Communication Specialist 2nd Class Justin Stumberg: 210; Photographer's Mate 2nd Class Michael B. Watkins: 109

Zuma Press: Andrew Davis Tucker/Augusta Chronicle: 233

FRONT and BACK COVER:
Top row from left to right: U.S. Coast Guard; U.S. Army/U.S. Geological Survey; Damon Winter/The New York Times/Redux

Middle row from left to right: U.S. Coast Guard; NASA; Getty Images

Bottom row from left to right: U.S. Geological Survey; Eric Thayer/Reuters; Justin Sullivan/Getty Images